ELOQUENCE OF THE SARDINE

ELOQUENCE

OF THE

SARDINE

Extraordinary Encounters Beneath the Sea

BILL FRANÇOIS

Translated by Antony Shugaar

ST. MARTIN'S
PRESS
NEW YORK

Library of Congress Cataloging-in-Publication Data

Names: François, Bill, author. | Shugaar, Antony, translator.
Title: Eloquence of the sardine : extraordinary encounters beneath
  the sea / Bill François ; translated by Antony Shugaar.
Description: First U.S. Edition. | New York : St. Martin's Press, [2021] |
  "Originally published in France in 2019 by Fayard"—T.p. verso.
Identifiers: LCCN 2021006909 | ISBN 9781250272430 (hardcover) |
  ISBN 9781250272447 (ebook)
Subjects: LCSH: Oceanography. | Aquatic animals. | François, Bill. |
  Marine scientists—Biography.
Classification: LCC GC11.2 .F73 2021 | DDC 578.77—dc23
LC record available at https://lccn.loc.gov/20210069

Originally published in France as *Éloquence de la sardine* in 2019 by Fayard

First U.S. Edition: 2021

10   9   8   7   6   5   4   3   2   1

*To my mother, who taught me the joy of
building worlds with words.*

*To Mickey Taylor, who painted for my eyes
the poetry of wild rivers.*

*To the fish of the Mediterranean, to all the
other fish, and to those who take delight in
learning about them.*

To you all, who will hand down their stories.

CONTENTS

Before 1

Any Fish Will Tell You So 8

The World Without Silence 18

Packed Like Sardines 29

Are Fish Good at School? 44

Cockles and Mussels 60

Daily Specials 81

Draw Me a Fish 93

Hold an Eel by the Tale 103

Sea Serpents 117

The Sea Is Your Mirror 130

Aquatic Dialogues 143

In Tune with the Tuna 159

The Tail End 172

Epilogue 179

ELOQUENCE OF THE SARDINE

BEFORE

The boulder was so high that I had to take off my beach shoes so as not to slip while I climbed. It was more comfortable that way. With their rusty buckles and translucent plastic straps, these jelly shoes—or "jellyfish" sandals, as we call them in French—hurt my feet even more than their undersea namesakes. And they slowed me down with each step I took in the water. I preferred to take my chances on the jagged rock edges, even if that meant spending the rest of the vacation with my ankles covered in water-resistant Disney-character Band-Aids.

I was determined to reach the top of the rock. This promontory stood at the end of the sandy beach, where grown-ups drowsed behind their books. On the near side of it, the merciless "summer workbook" awaited me; beyond it stretched the wild coastline. From the summit you could survey the entire small cove, the tide pools and channels running between the rocks. The sea swelled in and ebbed out like a slow respiration, and when it breathed in, the water was so smooth you could see, through its crystal-clear depths, all that hid beneath the surface. This was the perfect moment to observe the creatures who live under the sea. I loved seeking out those creatures, waiting for the sea to inhale so I could spot them, try to catch them in my net. All of them fascinated me: green crabs with seaweed wigs, translucent shrimps, periwinkles blowing streams of bubbles, and even the scarlet sea anemones I didn't dare touch because grown-ups had warned me that they stung. The only creatures I wanted to avoid at all costs were the fish that lived far from the rocks, out in the open water where my feet wouldn't touch bottom. Those fish scared me. My parents would sometimes bring them home from the market, and their big round eyes frightened me, as did those two slits just behind their head which made them look like decapitated beasts. For fear of those fish, I never dared venture beyond the world of tide pools and shore rocks. The free blue water that I glimpsed farther out awakened in me a profound terror.

From the height of that rock, as the sea inhaled, I glimpsed something glittering at the edge of the waves. A gleam that riveted my gaze. Perhaps it was a tiny treasure, a piece of pearly seashell or some lost object. I had to go see. Cautiously edging my way down the jagged rocks, I approached the flash of light. And in that moment I met my first sardine.

I didn't know then that it was a sardine, nor how rare it was to encounter one so close to shore. Normally sardines live in the open water. This one had probably lost its way, perhaps fleeing a tuna, also a rare occurrence because, back then, there weren't many tuna left in the Mediterranean Sea. Have you ever seen a live sardine? Few people know just how lovely a live sardine is. This one was shining and silvery, with an electric blue line like a garland along its back. On its flanks was a broad streak, glowing and golden. It was at once resplendent and fragile, like one of those tinplate collectible toys that so tempted me when I saw them in stores, but which I was forbidden to touch except "with my eyes." By the way it rolled on its side, tormented by the waves, I could tell this sardine was not in great shape. It didn't even seem bothered by my presence, whereas normally even the smallest shrimp would scurry off at the mere vibrations my feet sent through the water.

Carefully I collected the sardine in my net, then gazed in disbelief at this remarkable gift from the sea as it writhed in the water of my plastic pail. The sardine stared up at me with its black-and-white eye; it seemed to be trying to tell me something. I sensed that its silence contained secrets it wanted to confide in me, about its life in the deep blue sea, about its strange daily routine as a sardine. Its existence, the way it perceived the universe, intrigued me. I wondered what landscapes it moved through, what creatures it swam alongside, and whether it conversed on occasion with other sardines. All at once, the deep ocean ceased to frighten me; I felt attracted to the silent secrets of the sea.

Little did I suspect that, after this chance encounter with a sardine, my fascination with these undersea mysteries would stay with me. That I would be carried ever farther out to sea in search of an underwater universe whose captivating inhabitants, by no means silent, would willingly share their stories.

How do these creatures communicate? Through what senses do they experience the world? Are their lives and their emotions comparable to ours? Driven by the desire to solve these mysteries, I became a scientist. Hydrodynamics and biomechanics, my chosen fields of research, have offered me fresh insight into the marine world, revealing wondrous new answers, and even more new questions.

Since then I have swum, sailed, and even dove, by day and by night, to observe these fascinating creatures. Back when, for fear of fish, I didn't dare venture any deeper than where my jelly shoes could touch bottom, I would never have dreamed that soon I would spend my days studying them and my free time traveling to find them. I wouldn't have believed that one day I would hear whale song, visit the swordfish of the Mediterranean, count albatrosses, or play with manta rays. Nor that I'd find, just a short walk from my home in the middle of the city, even more remarkable fish and other aquatic creatures.

Along the way I've also encountered fellow humans who've bound their fate to that of the sea: scientists who shed light on its secrets, fishermen who live in harmony with the water, volunteers who devote their time to protecting it. I've shared in their projects to better understand the undersea world, to preserve it, or simply to find my place in that ecosystem and learn how to converse harmoniously with the ocean. They have taught me how to understand the signals of dolphins, how to catch tuna, how to approach seals. So I have come upon other stories, stories written or narrated by humans, illuminated by science or by the magic of legends, embellished by the innovation of discovery or by the poetry of oral tradition.

What have all these stories taught me? In addition to sharing its captivating beauty, the undersea world offers us other kinds of knowledge, especially about ourselves.

The inhabitants of the sea have, first and foremost, taught me to speak. Their ways of communicating, each species in its own fashion, and their ability to create stories in spite of the sea's apparent silence have revealed to me the art of rhetoric. These stunningly eloquent beings have confided their stories to me and given me the desire and the inspiration to recount them in my turn. It is thanks to these creatures that I am able to share with you, in this book, the stories they have led me to discover.

This book will plunge you into the depths of the ocean and of history, of the scientific world and the world of legends. I will introduce you to the secret society of anchovy shoals, and together we will take part in the conversations of whales. As we go, we'll meet some unusual characters, such as Åle the eel, who lived for a hundred and fifty years in a well, and the remora, friend to the Indigenous people of Australia. We'll take time to listen to the song of the scallop and the ancient saga of the extraordinary periwinkle. We'll decipher the latest scientific discoveries on the immunity of corals or the changing gender of wrasses, and be lulled into reveries by the ancient legends of mariners, often more believable than the incredible truth.

I hope you will emerge from your reading the same way I emerged from the water that first time I dove in it—with your head filled with stories and an immense desire to recount them in your turn. And I hope your beach vacations or visits to the aquarium will never be the same, and that you'll look in a new light at your goldfish, your seafood platters, and your tuna salad sandwiches.

The sardine flopped around in my pail, leaping up the sides decorated with pink and blue starfish. It was clearly yearning to be back in the sea. So I carried it to where the inlet flowed into the ocean, where the water was calmer and deeper. Balancing on the slippery rocks, taking care not to spill the pail, I made my way to a small stretch of shoreline and emptied it into the water, safe from the breakers.

As it swam away, hesitantly, toward the open ocean, the sardine motioned for me to follow it. It invited me to come along and began to recount its story.

How did the sardine tell me its tale? That will remain my secret. Everything that follows in this book is absolutely true. Whether my sources are the findings of rigorously peer-reviewed scientific studies, references from ancient works, or personal anecdotes and observations that plenty of witnesses can confirm, they are all trustworthy and verifiable. But as for the way that sardine began telling me its story, I'll have to ask you to take my word for it.

It was a long time ago, and I no longer remember it very well. And besides, how many good stories come to life without slightly strange beginnings? Let's simply follow along together, the way I did as a child. Together we'll listen to the sardine's tales, which have changed my way of seeing the oceanic world and have improved my understanding of our own.

That day, after returning from the beach, I spent the evening rummaging through the trunks in the garage for a diving mask and a snorkel. I was slightly afraid that I would swallow water through the snorkel, or that the mask, too big for me, would leak. Little did I realize, as I fastened the glass to my face, that I was on the threshold of a new world, and that I would never fully return to dry land.

ANY FISH WILL
TELL YOU SO

In which we dive under the waves to better understand what fish feel under the sea.

In which we wonder whether our ancestors learned to speak by diving.

In which we observe that, under the sea, color and scent are a language unto themselves.

In which we discover that the subtitles of the silent ocean can be read in invisible worlds.

The hardest part is going in up to your shoulders. As long as the water is only at your calves, or even your waist, you're still on dry land, you can still cling to the warmth of the sun. But once the water hits your shoulders, you always shiver. You're entering a hostile, all-enveloping chill. Nothing to do but take the plunge.

The first time I went diving in the sea, I let out a strange cry at the nip of the cold water.

Since my face was covered with a mask, only a raucous trumpeting escaped the snorkel to express my astonishment at this simple yet surprising discovery, a sort of prehistoric grunt, perhaps meaning "wow, that's cold!" With a plastic tube in my mouth and a Plexiglas screen over my eyes, this world, so blurry beneath its reflective surface, was suddenly unveiled, crisp and crystal clear. Once I was over its thin border, this inhospitable element suddenly became transparent and buoyed me gently. I could fly, look, and breathe through it. But I couldn't speak. The snorkel had transformed my voice into bursts of crude, primordial breathing sounds. I sent words into the tube, but only animal sounds came out. It was like a sort of tacit pact with the elements. I had gained the power to see what the sea concealed, to fill my ears with its sounds, to glide, cradled by its weightlessness, but I had lost the ability to express myself in words.

It was a strange sensation, something between setting out to conquer a new universe and relapsing into a primitive state, back to mankind's distant origins, long before humans had gained mastery of language.

As a medium, water is both hostile and welcoming to human beings. We're afraid to dive in, but we are in fact perfectly equipped for the plunge. Our bodies are amazingly well adapted to it. It takes only a splash of cold water to the face for our diving reflex to automatically and immediately drop our heartbeat by about 20 percent, preparing us to hold our breath and go under.

The human body is equipped for aquatic life in many ways, perhaps even too many for it to be a mere coincidence. Hence some anthropologists have postulated that our ancestors evolved differently from apes specifically in order to enter the water, eventually becoming human beings.

How else to explain our lack of fur, our subcutaneous layer of fat, unlike anything found in other primates, or the millions of enormous sebaceous glands that oil our skin, those efficient lubricating organs that no other land animal possesses in such quantity and size? All these mysterious and seemingly useless characteristics of our bodies that set us apart from apes could, therefore, actually be adaptations to the aquatic environment. Our skin, like that of marine mammals, is smooth and hairless; it is lubricated by sebum, making it waterproof, and our fat pads insulate us against the cold. Another curious fact: a human newborn can hold its breath under water by reflex action and float on its back, whereas a baby chimpanzee would sink and drown. Two million years ago, when our genus broke away from that of the future chimpanzees, our ancestors probably had to find food on the seashore or in wetlands to survive the arid savannas. They would have stood up on their hind legs to better keep their footing. And while diving for crustaceans, roots, or stems of water lilies, they would have learned to control their breathing, until, through the process of evolution, the larynx descended and the vocal cords developed. So diving and swimming arguably gave us the foundations for two abilities key to our evolution: bipedalism and speech.

Just how much credence should we give this idea? The hypothesis, popularized in the sixties, spurred distrust and controversy. The idea that our evolution should have passed through a "missing link" to an exclusively marine way of life sounds like a baseless exaggeration. But recent studies of southern African fossils suggest that aquatic environments played a critical role in human evolution, sometime around 2.5 to 1.5 million years ago. To survive the dry season, our ancestors were forced to adapt to water-based environments in order to find food in the oases. This adaptation helped drive early humans to leave the shelter of forest trees to try their luck in the plains, and eventually conquer the rest of the world.

We may have come down from the trees, but we never truly conquered the sea. In the water, as any fish will tell you, we can't see everything. For seeing alone isn't enough.

During my very first dives down to the rocky Mediterranean seabed, I was amazed to discover the sheer diversity of marine life. I watched wrasses and sea breams glide above the mossy rocks, and I was captivated, as though watching a performance. As on a stage, there were images full of light and movement, and there were sounds, mysteriously rumbling and popping in my ears. I thought I was privy to the whole of the show, but in fact, I was enjoying just a part of it. I was watching a silent film, without its subtitles, unaware of the countless conversations taking place outside the reach of my eyes and ears.

The subtitles of the sea are written in, among other things, the language of scent. In the oceans, smells and perfumes are a language unto themselves. Water is full of scents we can't detect. When we go underwater, we often pinch our noses, unless our diving mask does so for us, making it far preferable to swimming goggles. And rightly so, for as unpleasant as it is to swallow seawater, it's far worse to take it through the nose. The fragrances of the sea remain inaccessible to us.

Yet the currents carry countless odorous molecules. Fish can smell them; they live in a galaxy of aromas. They can distinguish very slight nuances of smell, distant and infinitesimal odors, in the water. Some odors stick to our memories: the scent of places, old books, seasons, or people. The scented reminiscence of such things reawakens imprinted feelings. Fishes' memories brim with such olfactory recollections.

From the waters off Greenland, the Atlantic salmon can detect the smell of the Breton stream in which it was born, then follow its traces for thousands of miles in search of the mouth of its native river. This memory is many years old, created back when the salmon was just a small fry, surfacing in the summer to take in the scents of evening while filling its swim bladder with air. This memory is a tiny concentration of odorous molecules. A few drops in an entire ocean come from that stream, diluted among the immense mass of countless drops from countless other streams. But the salmon recognizes them and always manages to find their source.

Fragrances evoke so many emotions that fish use them to communicate. While we see nothing but fish swimming, the water all around them is filled with invisible plumes of emotionally charged perfumes. These are pheromones, chemical translations of their moods. Scents of stress, love, hunger, and more. The scents are targeted, but they are occasionally intercepted by unwelcome sets of nostrils. A small fish's distress scent warns its kin of danger, but it can also be a mouthwatering signal for predatory fish. Damselfish, small colorful tropical fish found in coral lagoons, use this flaw in their communication to their advantage. When one is wounded and captured by a predator, it emits extra S.O.S. molecules to attract even more predators! Inevitably the predators start to fight over their prey, and amid the confusion the damselfish takes advantage of the confusion to seize its opportunity and makes its escape.

Snorkeling over the marine depths feels like being riveted to the sky. You discover a universe as you glide above it. The farther you move away from the shore, the deeper the water gets beneath you. And as the seabed drops away, the colors on the bottom, ever more distant, are tinged increasingly with blue, until everything is steeped in the same shade of faded ink. This thin blue also conceals another subtitle of the underwater spectacle: its invisible colors.

Water makes colors disappear. Sunlight contains the wavelengths of all the colors we know on the surface. But the farther light travels through water, the more some of these colors are absorbed by the water molecules it encounters there. Water molecules greedily devour colors. First and foremost, they guzzle down the "warmest" colors, the ones with the longest wavelengths—red, orange, yellow. At a depth of five meters, all the red vanishes. Red objects melt into the blue, and we can no longer detect their hue. As the light continues downward, it progressively loses yellow (at fifteen meters), then green (at around thirty meters). Soon all that remains visible is blue. Below sixty meters the sea becomes a monochrome azure to our eyes. Then even the blue disappears, and we enter the abyssal dark. At four hundred meters there is no more sunlight at all. Only luminescent creatures brighten the darkness. But when a beam of light plunges into the depths, it contains other, invisible rays— the ultraviolets, those colors that are "bluer than blue." Our eyes are unable to perceive these very short wavelengths, because our lenses block them. Fish, on the other hand, are extremely sensitive to these colors, which illuminate their world in places where we would see nothing but blue. Some underwater landscapes and animals may seem dull to us, but if we observed them with an ultraviolet-sensitive device we would be stunned to discover their multicolored patterns, their countless vibrant patches and stripes.

Under the sea color is yet another language of its own. Many species can change color on command as a form of communication, doing so even more effectively than chameleons. If you look at a fish's skin through a magnifying glass, you'll see an array of tiny dots in various hues. These are chromatophores, pigmented cells that a fish can dilate or contract at will to varying degrees, depending on the species. By choosing which chromatophore to dilate, the fish can choose which color to display, as if selecting its own screen pixels.

It can even change the patterns of its skin in the same way, so that they become yet another signal that a fish can use to express itself and converse with others. It's a form of communication so subtle that it remains largely a mystery to us.

Color transmits information, but it can also convey lies or illusions. There are true colors, like those of salmon eyes, which serve to express moods and emotions. But there are also the eye spots of the wrasse, which deceive by imitating the eyes of a predator. There are the polarized signals of the mantis shrimp, which they alone can decipher, coded on their shells exactly the same way that 3-D films are coded for polarized 3-D glasses. There are the stripes of the marlin, whose ultraviolet color is precisely calibrated to the wavelength that best blinds the mackerel's eye. The marlin uses these stripes to convey its moods to its peers, but also to stun entire shoals of mackerel with dazzling signals that they can't understand. Seized with panic, the mackerel gather into tight balls, which the marlin can easily lambaste with its swordlike bill.

Even less perceptible to us than the colors and scents of the undersea world, found somewhere between the boundaries of smell and hue, other subtitles of the sea are written in languages that defy our imagination.

For instance, I couldn't tell you what a vortex feels like. All the signals conveyed by the currents and vibrations of the water are indiscernible to us humans, but not to fish, who detect these motions via their lateral line and leave similar whirlpools in their wake, just as an airplane writes its signature in white contrails in the sky. Fishes' lateral lines are covered with hair cells, whose tiny cilia bend under the effect of

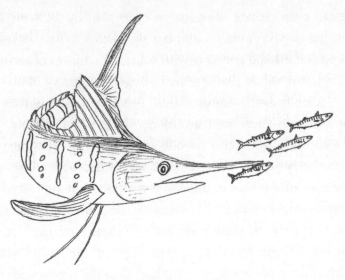

MARLIN AND MACKEREL

the currents, transmitting information to the nervous system. Fish can in this way chart the flow of water around them. Decoding these eddies and currents allows fish to find their way through total darkness. They visualize an image of their environment through the currents and movements of the water, a map that overlays other images comprising colors, sounds, and scents—a reading of the world which we can only imagine in dreams.

Nor could I describe, even to myself, the world of electric fields, those impalpable forces that certain fishes like the electric ray pick up and use to send signals and locate their prey. It's like a whole second ocean, in another dimension, where each living being has an imprint, an appearance, and a voice of its own. When night reaches the depths of the reefs, sharks tap into this second universe, and they use it to hunt and find their way. How exactly does it look like to them? That mystery remains their silent secret.

Many more of these parallel universes remain a matter for speculation. For instance, certain fish are thought to perceive magnetic fields. Migrating fish would use this ability like a built-in compass to orient themselves, adding another layer of subtitles to the immense, imperceptible spectacle of the sea. Magnetic sensing is yet another way to navigate the vast blue space, a reading of the world altered by the signals that the Earth itself sends out like an immense magnet.

But we should not envy the underwater creatures their various ways of communicating. We, too, have many means of conversing. By speech, by writing, by gestures, images, symbols, music. These are also hidden parallel worlds offered up to our senses. In fact we often complain that we have too many communication channels, for instance when someone we contacted by text message replies by email, then begins conversations on multiple social media at the same time, only to ignore us when we call them on the phone. Ocean dwellers similarly initiate their conversations over a thousand invisible networks at once. Their stories are carried along on a wide array of waves and channels—imperceptible colors, electromagnetic fields, vibrations in the water, pheromones. But they also converse the old-fashioned way, by their equivalent of the rotary phone. Or perhaps even something preceding the telephone: by simply talking to each other.

So let's listen in.

THE WORLD
WITHOUT SILENCE

In which distant volcanoes and invisible whales sing in the gurgling of our waterlogged ears.

In which the seahorse's xylophone has a purpose other than racking up Scrabble points.

In which the rock lobster plays its violin out of tune.

In which we let ourselves be lulled by the scallop's song.

The first time you plunged your head under the sea, you heard a strange sound. An acoustic pandemonium, both rumbling and clicking, as if you couldn't hear clearly. And when you lifted your head back above the surface, water dripping from your ears, you may have thought to yourself that you hadn't heard anything, that human ears must not be adapted for listening underwater, that the cacophony was merely an illusion.

The truth is that our ears do work perfectly well underwater. You had just heard the voice of the sea, and its very first story—a story that is a blend of all its other stories at once.

The sea is full of sounds, even more so than the air we inhabit. Sound is simply a vibration of matter. Water, which is denser than air, vibrates better, and therefore transports sound better. Underwater, sound travels farther than light, crossing miles and miles without fading. The voices of the sea contain sounds from far away, signals whose sources are impossible to see. Noises we'd never dream of hearing from the beach, noises that connect us to their remote origins.

These gurglings that haunt our waterlogged ears are part of a churning soup of noise. Multiple voices blend together like shredded vegetables in a stew. As with the subtle nuances of a perfume, different notes can suddenly stand out clearly, only to fade into the background again. Like the various instruments of an orchestra, each voice of the sea has its own distinctive note and wavelength, and sings its story in its own timbre.

It is the mixture of these timbres that creates this unsettling noise, flooding our ears without drowning them. Acoustical oceanographers call it "underwater ambient noise."

Let's listen to it.

First you hear the bass notes. Underwater, the background noise is a low-pitched sound. It rumbles and thunders like a kind of snore. This noise, the most intense sound in the sea, is the echo of various elements: waves crashing onto the shore, winds sweeping the surface, but also the Earth and all its whims. You can hear in it the cracking of the polar icebergs, the grinding of earthquakes along the ocean ridges, the gust of distant storms. . . .

The rumble of these cataclysms arrives from far away, low and exhausted by the journey, and makes up the backdrop ground noise for the marine orchestra.

You can also hear a sort of fizzing sound, like the clacking of maracas. This is the sound of rain, the bubbles of foam on the water's surface, where the gaseous and liquid phases meet.

Above these low notes you can make out long violin vibratos reverberating for dozens of miles. These are the grinding sounds of ships' engines, the screech of metal, the hissing of screws. Maritime routes are as noisy as our highways, but can be heard from much farther away. A passing container ship makes as much noise underwater as an airplane taking off does in the air, and traffic at sea generates background noise every bit as intense as a busy street.

Other, more melodious voices try in vain to be heard through this racket. If you listen closely, you can hear sounds like whistling flutes or tooting trumpets. These are the echoes of whale song.

This music brims with meaning that scientists are only just beginning to decipher. There are love songs, lullabies to soothe whale calves, ditties to celebrate a feast of herring. Some melodies, it seems, are even sung for the sheer pleasure of the music.

Although whale songs are seldom clearly distinguishable to the naked ear, they account for a substantial portion of the sea's underwater ambient noise, and can be heard throughout the entire ocean. That's because whales, eager to talk to each other across the vast expanse of the sea, know how to make themselves heard from very far away. They've developed their own underwater long-distance telephone network.

The whales' telephone package deal simply works by means of pressure and temperature. There are two layers of water in the sea: surface waters heated by the sun, and deeper, colder waters. At the thermocline, the interface between these two zones, the temperature drops abruptly. You may have unknowingly noticed this when your foot dipped into one of these "cold currents," near the seabed, while you were out for a swim. In the high seas the phenomenon is even more pronounced. Water temperature suddenly drops by almost forty degrees in a few dozen fathoms of depth.

This boundary between hot and cold water traps sound. Sound rising toward the surface will ricochet off the warmer waters, where the elevated temperature accelerates its propagation and bends its trajectory back into the depths. As it descends, sound rebounds off the deeper waters, which are at a much higher pressure, and is further accelerated and sent back up toward the surface. Sound is thus imprisoned between masses of water along the thermocline. When whales sing directly into this sound canal at the boundary between cold and warm waters, their voices ricochet along the thermo-cline and therefore travel outward in a straight line, without straying or fading, for thousands of miles, in exactly the same way that light travels when trapped inside a fiber-optic cable.

The fin whales of the Mediterranean use this telephone network, known as the SOFAR or deep sound channel, to serenade each other and arrange meetings from distances of more than 1,200 miles.

You have to be lucky and in the right place to clearly hear the melodies of whale song. But these sounds are mixed into the general churn of underwater ambient noise, and you can hear their notes in any ocean simply by plunging your head underwater. Scrutinizing the ocean's noise and analyzing its individual notes, in fact, is a method used by cetologists (cetacean biologists) to study the rarest whale populations and to listen to animals that cannot always be observed directly. Each species has its own voice and its own wavelength, like the FM or AM channel of a radio network dedicated to its conversations.

In 1989, in the Pacific Ocean, hydrophones first captured the call of the world's loneliest whale. A whale was emitting songs that seemed characteristic of fin whales, but at a frequency of 52 Hz, a sound equivalent to the lowest note of a tuba. That frequency was far too high for fin whales, who communicate at frequencies between 10 and 35 Hz. As a result, this whale has been singing, speaking, and calling out to its fellows for decades without a reply. It wanders alone through the vast expanse of the open seas, and year after year only oceanographic hydrophones hear its calls. We have no idea why this whale has such a strange voice. Some believe it's a hybrid between a blue whale and a fin whale. Others believe it's a birth defect, or that this whale was born deaf and could never correct the sound of its voice. Nor do we have any idea whether, in the immense expanse of the seas, this whale has ever crossed paths with other whales, much less what it felt when it saw them without being able to speak to them. No one has ever laid eyes on this whale, even though we've been able to track its solitary yearly migration by listening. For human beings this lone whale exists only as a song, the very song that isolates it from its peers, a song it sends out tirelessly, full of hope, into the void of the Pacific.

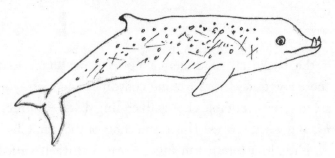

BEAKED WHALE

In the Atlantic there's a whole species of whales that produce unidentified sounds, without anyone ever having been able to observe or describe a single specimen. Analysis of the sound's structure suggests that these cetaceans belong to the family of beaked whales, extremely shy animals that never blow plumes of mist on the surface, and they are quick to dive whenever boats draw near. This is a strange family of cetaceans. The rare beaked whale species that have been fleetingly observed have long brown mottled bodies, and tusks similar to a wild boar's. They hunt squid down to depths of 9,500 feet, a record for marine mammals. Ultimately, the voice of these discreet animals is what gave us the most information about their behavior, allowing us to know them better, and even to discover a new species of beaked whale, which, for the time being, remains unseen! The oceans abound with such hidden stories from timid animals who are nevertheless eager to tell their tales. The depths of the sea are full of shy creatures whose solitary sounds conceal the wonders they don't dare share face-to-face.

Against the backdrop of undersea noise, some songs soar above these low tones. Approaching coastlines and reefs, you will start to hear a veritable chorus, blending diverse timbres, rhythms, and vocal ranges. This is the choir of the fish. Chattier than birds in a forest, fish fill the seas with their varied cheeps, each species with its own unique sound.

Some fish emit sounds through their swim bladders, pouches of gas located in their abdomens that keep them neutrally buoyant. They use these swim bladders as a drum, like children who tap out rhythms on their bellies after eating, an unaccountable music we've all made at one time or another. Patting their bellies like this, with the help of special stomach muscles, drum fish croak, groupers grunt, and gurnards rumble. Their sounds are reminiscent of fog horns, drum so-los, or TV game-show buzzers. Some of these fish can even be heard from shore, while others murmur discreetly. The cod is more talkative than the haddock or the hake; the drum fish has a deeper song than the perch.

But this is not the only way fish can sing. The jack (or tre-vally) and the sunfish prefer higher notes and grind their teeth to play screeching melodies. The seahorse plays its own per-sonal xylophone by scratching its neck with the bony ridges on the back of its head, while the catfish makes a high-pitched noise by plucking its spines. As for the humble goby, found in tide pools, no one has yet managed to figure out the hydrody-namic mechanism that enables it to sing its love songs simply by blowing water out its gills.

When day breaks over the most densely populated coral reefs, the fish choir reaches the sound level of a cocktail bar at happy hour. Even this pales in comparison with the Gulf weakfish, whose song grows louder than 200 decibels when it gathers in shoals to spawn in the Gulf of Mexico, temporarily deafening nearby cetaceans.

If you listen carefully to the noise of the sea along our coastlines, you'll notice crackling clicks rising above the rest of the soundscape, like bursts of percussion—multiple, rhyth-mic, vibrant tappings, rising above the rest of the orchestra.

These are the soloists of the sea.

The clicks come from shellfish snapping shut, sea urchins grazing on the rocks so that their shells echo the sound of their teeth, prawns snapping their pincers . . .

Mantis shrimp can snap their claws so hard and fast that

the resulting shock boils the surrounding water by cavitation. The sudden pressure decrease creates a vacuum bubble, which implodes with a bang. It sounds like a gunshot and is the loudest noise heard below sea level. The shrimp use this ability not only for self-defense but also to capture food, stunning their prey with sonic shockwaves before moving in for the kill. Rock lobsters, meanwhile, consider themselves more musical. They play the violin with their antennae. Using the same friction-based mechanism as a violin bow, which clings to the strings as it slides over them to produce a vibration, they rub their antennae against the base of their eyes. Their shells amplify the sound produced, much like a violin's sound box. These are the only living things (besides human musicians) known to produce sounds based on this friction principle. Alas, rock lobsters are not rock stars. They play rather off-key, hitting lots of wrong notes, and the noises they emit sound more like a squeaky hinge. They make use of this unbearable racket to keep their predators at bay.

Sometimes the clickings of the sea merge into a vaster melody. Scallops are timid creatures, especially in the presence of octopuses or starfish, who would eagerly make a meal of them. They tirelessly scan the undersea horizon with their rows of blue and black eyes, for scallops have eyes, a rare luxury among shellfish. At the slightest hint of trouble, a scallop will make its escape by opening and closing its shell rapidly to propel itself into open water and swim away by clacking its shell in the same manner. The scallop also discharges waste water and irritating sand particles in a kind of submarine sneeze. These clackings and sneezings form part of the underwater sonic landscape of the scallop habitat. In Saint-Brieuc Bay in Brittany, they create a veritable concert of castanets and coughing. Although scallops don't use this sound to talk to one another, they communicate plenty to us. By listening to their song and studying the frequency of their clackings, we can learn whether their water is pure or polluted, and how numerous their predators are. They provide oceanographers with information about the state of the sea and the biological health of their environment. With their concert of sneezes, scallops illuminate for scientists some of the mysteries of their strange lives.

The background noise of the seas also contains sounds that humans simply cannot hear. At one end of the spectrum, infrasounds are too low-pitched for us to detect. The noises produced by the displacement of huge masses of water, the movement of fish, and the eternal turbulence of the waves are all beyond our hearing abilities. At the opposite extreme are ultrasounds, murmurs too high-pitched for us to hear, such as the clicks of dolphins' sonar or the noise produced by the thermal agitation of water molecules. With these sounds we get close to the limits of what constitutes sound itself, since this thermal noise is generated by particles of that same medium that transports sounds. It is almost a theoretical sound, surely the ocean's most intimate and secret voice. It is the noise of water itself—not the noise of its movements, its inhabitants, or even its currents, but the noise of its very molecules, water's actual matter, its existence. This music is hard to imagine. What would the water say to us, what would its voice sound like if we could hear it? Science tells us it's a perfectly random white noise, all the louder for being sharp. This doesn't make it any easier to conceptualize.

Dolphins can hear this ultimate sound, and for them it's simply part of the sea's background noise. They likely find it rather annoying, as it disrupts their sonar signals. But who knows, perhaps they can also decipher some of the ocean's secrets in that impenetrable song.

Underwater ambient noise is a mishmash of sounds, a mother lode of noise, in which all the dissolved voices of myriad invisible beings share their stories. From the storm to the water molecule, the blue whale to the shrimp, each plays its own instrument in the orchestra, tossing its own sprinkling of notes into the mix.

Science and our imaginations try to give meaning to this hubbub, offering various interpretations of such a confused and wonderful dream. What a delightfully dizzying idea, to think that we can hear the echo of all those stories and all those voices speaking to us through the gurgling of our waterlogged ears!

It's our good fortune to hear them, and a miracle to listen.

Let's explore these secret stories together.

PACKED LIKE
SARDINES

In which the sardine makes itself into a mirror of
the ocean.

In which herrings fart up a storm.

In which the wrasse offers free shaves.

In which we are no more ourselves than corals are.

Before you see a shoal of sardines under water, you first
glimpse only furtive flashes that catch fleeting beams of
sunlight. Even when present in large numbers and close at
hand, sardines know how to remain invisible. Their backs are
blue as the sea; from above you can't see them at all. Viewed
from below, their pearly bellies disappear in the light of the sky.
Seen from the side, their flanks are as reflective as mirrors. In

the blue of the water, sardines take on the surrounding color and blend into the seascape, becoming nothing more than a blue reflection, the very picture of their environment.

It is the stratum argenteum, a layer of skin located just below the transparent scales of many species of fish, that lends them their silvery appearance. This brilliant skin is much more than a simple mirror. It reflects light much better than even the most perfect looking glass. On shiny materials such as mirrors, light is more or less strongly reflected as a function of its angle of incidence. Reflections are not equally strong in all directions. This is due to a fundamental and invisible property of light, its polarization. The ray of light reflected off a material is polarized, meaning its electric field vibrates in specific directions, in accordance with the electron's vibrations on the reflective material. The light can therefore be reflected only if it hits the surface at specific angles. Thus, a shiny object reflects irregularly, making it possible to distinguish it from its surroundings because it gleams only from certain parts. And because these reflections are made up of polarized light, they are blocked by the filters on polarized sunglasses, which remove reflection and glare thanks to this phenomenon.

But the reflection of light off the sardine's skin is another story. That skin contains reflective guanine crystals with two different shapes, each of which polarizes light from a different angle. Therefore, no matter what direction the light shines in, it is always perfectly reflected by one or the other set of crystals in the stratum argenteum. A sardine is a perfect mirror, reflecting homogeneously from all angles. It is capable of blending completely into the universe it reflects off its skin. No eye can distinguish the sea itself from its reflection on a sardine.

On top of that silvery skin, as a form of protection, sardines, like most fish, have scales. Scales tell a fish's personal story. They are made up of concentric rings, very much like those in a tree trunk, that keep pace with the growth of the fish itself. Each ring represents an episode in its life. Tightly spaced rings indicate a harsh winter; more widely spaced rings point to faster growth in a generous summer. Some rings recall the mating season, or perhaps, for a migratory species, a crossing of the threshold between sea and fresh water. A fish carries a résumé of its life on its scales. If one is torn off, another starts to grow in its place. It simply resumes the story where it left off, writing the sequel without re-transcribing the past.

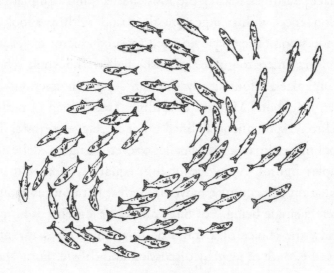

A SHOAL OF SARDINES

If you've ever observed a sardine carefully, you'll have noticed that it is not uniformly silver. It often displays a row of faint black spots behind its head and down its side. These small spots are markers that allow sardines within a shoal to identify each other more easily, the better to coordinate their swimming. There are fifteen or so individual sardines per cubic meter of a shoal. Relative to their size, that is close to the density of humans packed into a subway car at rush hour. However, unlike the passengers in a subway car, a sardine never swims against traffic, never bumps into its neighbors, and never causes even the slightest disruption or congestion. Without having to speak, sardines maintain respectful distances and speeds.

Simply by keeping an eye on its nearest neighbors and listening to the flows of water in their wakes, the sardine adjusts its pace. Sardines master the most perfect and complete art of eloquence: with one gesture all is said, with one look all is understood. They have no need for a conductor or for any marching orders to guide their fishy ballet. The whole school self-organizes automatically through simple interactions between neighbors. Millions of sardines swim along in perfect synchrony, spontaneously lined up or in staggered rows. The school moves in an aquatic ballet of ever-changing, delicately complex figures. A myriad of fish, equal in number to the population of a country, moves in perfect harmony as though it were a single being. For our families or a group of human friends, the choice of a travel destination or a restaurant is often the cause of lengthy discussion and debate. For a shoal of millions of sardines, decisions instead are made naturally, without the slightest deliberation. If a predator shows up, the shoal adopts evasive maneuvers, splitting in two like a living fountain to disconcert their attacker. If a mass of copepods, small planktonic crustaceans on which sardines feed, happens to pass nearby, the shoal chooses the most efficient strategy to feed all its members based on the needs of the group. It might decide to disorganize itself so that each sardine can take advantage of the bounty individually, or instead it might just choose to advance in serried ranks, to devour the prey with systematic efficiency. A collective intelligence emerges from the sum of the small combined actions of each sardine. It's a fantastic form of democracy. Without a leader or a dominant group, without receiving orders, all the sardines in the shoal agree to swim together, even when the shoal is several miles long.

Herring, close cousins of sardines, also live in shoals, but they lack the sardine's good manners. When the time comes to regroup at nightfall, they have their own technique for communicating in the darkness. This is how they keep from losing touch with the group. It's a fairly crude technique, and one that once came close to starting a war.

In 1982, in the tension of the late Cold War and one year after the accidental beaching of a Russian submarine near

Stockholm, the Swedish navy feared a Soviet invasion more than ever. The Swedish press was constantly reporting signs of an imminent invasion. And then the "golden ears" of the Swedish navy, petty officers responsible for analyzing the sounds captured by sonar arrays, detected an unfamiliar and unexplained auditory signal. This "typical noise" appeared at the same frequency as the sound of a ship's engine propellers.

The Swedish military, believing they had just detected an ambush by Russian submarines, rushed to investigate. Several submarines were deployed to the sector, but it proved impossible to establish radio contact with the supposed invader, nor could it be observed on sonar equipment. Now convinced that they were dealing with an enemy in possession of powerful camouflaging technology, the Swedes sent out military aircraft and ships to patrol the zone for the entire month that followed. All units reported the same observations: they saw air bubbles rising to the surface where the signal had been picked up, but the submarine itself remained undetectable. Sweden narrowly averted a diplomatic incident with the USSR, which naturally denied any submarine presence in Baltic waters. Over the course of the following months, then years, the intelligence file concerning these "typical sounds" was reopened on numerous occasions. Each time the sounds were heard again, military officials and diplomats tried in vain to find out what was happening and defuse the situation. For the Swedish navy the insolence and agility with which these Russian submarines were taunting them was a serious affront. But despite these military efforts the unsettling noises continued to sow panic both in sonar rooms and in the halls of diplomacy. This went on long after the fall of the Soviet Union. In 1994 the Swedish government, at its wit's end over this ongoing affair, finally threw up its hands. Prime Minister Carl Bildt wrote a letter to Russian president Boris Yeltsin, upbraiding him for his inability to control the deployment of his fleet of submarines. Of course Yeltsin denied everything. It was not until 1996 that the Swedish armed forces authorized civilian experts, a team of bioacousticians working under Professor Magnus Wahlberg, to listen to the mysterious sounds, classified top secret, in an attempt to identify them. By analyzing the "typical sounds," the scientists absolved Russian submarines of all involvement and identified the real guilty party: a school of herring.

When they gathered for the night, these herring engaged in a rather unique form of chitchat. They communicated among themselves by means of flatulence! Their swim bladders, an organ that allows them to maintain their balance and control their buoyancy, is equipped with a complicated plumbing system that produces gas and expels it through natural channels. This concert of farts conveys very complex information, structured in rhythmic repetitions of sonic pulses every 32 to 133 milliseconds. The fish use these pulses to communicate in a range of frequencies that escapes the ears of their predators—except the Swedish navy. These farts also generate a poetic curtain of bubbles around the school of herring, which encourages them to remain clustered in darkness. The pearls of air rising amid their reflections form a harmonious nightly spectacle, one far more pleasant than the war they came so close to igniting would have been.

Fish communities aren't limited to vast, uniform shoals of sardines or herring. In the sea fish also establish social bonds among different species and invent languages in order to understand one another.

When night falls on the coral reefs, groupers and moray eels cooperate to hunt together. It's a fable worthy of Aesop's "The Fox and the Stork," the famous bedtime story. The Fox invites the Stork to eat with him and serves soup in a bowl, putting it out of reach for the long-billed bird. The Stork then invites the Fox to dinner in return and serves the food in a narrow-necked vessel inaccessible to the Fox, thus securing his revenge. Under the sea the Grouper and the Moray Eel, instead of teaching each other life lessons, collaborate to make any prey accessible to both of them. The grouper is an expert at sudden accelerations in open water and is equipped with a sharp eye, but it is not particularly agile when maneuvering. Its neighbor, the moray eel, on the other hand, can flush out prey by worming its way into their hideouts, but it's slow and can't see very well. So the hungry grouper pays a visit to the moray eel and sends him a special signal with its fins. The two of them then set off, side by side, in search of small reef fish. Once the grouper spots a hidden prey, it points it out with its nose, floating vertically, and the moray eel slips into the coral to flush it out. The fish that finds itself caught in their cross-hairs can neither flee to the open water nor conceal itself in the crannies of the coral. It can choose only which of the two predators will snap it up.

So this partnership is limited by the appetites of the accomplices. Whichever of the two, the grouper or the moray eel, is the first to capture its prey will swallow it whole in a single gulp. In other words, they share the effort but not the reward.

If you move in a bit closer to the rocks of the Mediterranean shores, you'll sometimes see fish of various species hovering vertically, motionless in the water, flapping their fins in a strange manner. It took me several days to summon the patience to observe this scene in its entirety, and to finally realize that these fish were waiting at a cleaning station for their appointment with a blacktail wrasse. This is a small fish of a violet black hue that rids other fish of their parasites, dead skin, and food remnants, which it then feeds on. By floating vertically in front of the rock, the fish signals its desire to be cleaned. Extending its fins as the blacktail wrasse approaches authorizes the wrasse to come do its job, which includes cleaning vital and sensitive organs like the gills. The cleaning station is a meeting place for fish, not unlike a beauty parlor or barber shop. All fish come in peace. Even the largest predators refrain from attacking the cleaning fish or any other prey in this zone of détente. There is often a long queue in front of the cleaning fish's rock.

Various species of cleaning fish are found all over the world, and the cleaner wrasse of the tropics has even developed business tactics. It can distinguish regular customers from drop-in new clients that it has never seen before. To encourage customer loyalty, if there's a line the wrasse will give priority to newcomers and to those who haven't come for a cleaning in some time. By so doing, it increases its regular customer base. Some in the hospitality industry would do well to take note.

But, as in all professions, the cleaning fish include a number of crooks as well. In the reefs of the western Indian Ocean, evolutionary complexity has allowed the mimetic blenny to imitate the cleaner wrasse, decking itself out in the same blue with black stripes as the authentic cleaner. But the imitator is not qualified to clean, quite the contrary. This con artist strips bits of skin and fins off its unsuspecting customers, then gobbles them up. In regions where the blenny operates, fish are far more mistrustful of the cleaner wrasse, who becomes, as a result, an even better merchant.

The undersea environment is an immense community that rivals our own cities in its diversity and complexity. Many different types of creatures live there together, each with its own role to play. The survival of a species often depends on the aid of multiple other creatures.

The coral itself, often the founding architecture of these communities, best personifies these support systems. It is the product of a partnership between the animal, vegetable, and mineral kingdoms. A branch of coral is composed of a multitude of tiny individual animals, polyps that resemble tiny sea anemones and live as a community. Their mineralized skeletons form the coral limestone, the original source of the white sand of the tropics. These tiny cnidarians feed in three peculiar ways. They can collect plankton with their tiny tentacles; they can devour polyps of nearby coral by projecting their stomachs onto them, and they can employ their preferred and most peaceful technique, gardening. The coral polyp cultivates inside itself a garden of microscopic single-celled algae called zooxanthellae. In exchange for shelter, a well-lit living space favorable to photosynthesis, and some nitrogenous waste as fertilizer, the algae provide the polyp with oxygen and food. It is this symbiosis between plant and animal that allows the coral reefs to grow.

But the coral's symbiotic relationships don't stop there. Marine biologists are now discovering that coral maintains numerous close-knit support systems with an array of other creatures, all indispensable to its existence. Coral has an acquired immunity to disease. It can protect itself from infections to which it has already been exposed, therefore resisting them better. And yet the polyp has no antibodies and no immune system like a human's. The current leading hypothesis to explain this disease resistance is probiotics. Coral's immune memory might be maintained by a population of various bacteria that live inside the polyp, just as we harbor an array of intestinal flora. These bacteria live in symbiosis with the polyp and defend it against external pathogens. They can then "remember" these pathogens, adapting to their attacks more efficiently.

In April 2019 a series of genomic and microscopic studies brought to light another family of inhabitants of the coral polyps, never before observed, the corallicolids. The role played by these creatures remains a mystery, but it has already prompted major speculation. They are present in the gastric cavities of polyps in 70 percent of coral species. They belong to the apicomplexan family, which primarily includes fearsome parasites such as those responsible for malaria or toxoplasmosis. But unlike their parasite cousins, corallicolids seem to live harmoniously with the coral, and they possess the genes necessary to produce chlorophyll, even though they don't actually photosynthesize. They seem to be situated halfway along the evolutionary pathway between plants and parasites. There is surely an unknown facet of the coral's life tucked away in this mysterious cohabitation, a story of mutual partnership and friendship. A microscopic friendship, but one that lies deep in the heart of the sea's functioning.

The coral polyp is certainly not an isolated being; it is impossible to separate it from the other microscopic species living inside it. This union has led to the construction of the most incredible underwater cities, cities that are visible even from outer space—coral islands and the Great Barrier Reef, where other creatures, from tiny prawns to large sharks, organize their own complex communities.

We have neither tentacles nor limestone exoskeletons, but are we really so different from coral? We also live in complex societies, where each link in the chain is meaningless without the others. Our civilizations and cities were founded on the fundamental principle of mutual aid, a principle we tend to forget in an age when individualism is lauded as an ideal. And yet our very bodies, like the bodies of many other animals, resemble the structure of coral. We house inside us immense communities of microscopic living beings that are not Homo sapiens, but whose fate is bound to ours. We are filled with bacteria, from our digestive systems to our mouths and down to the soles of our feet, and these bacteria are indispensable to our survival. The human body is thought to contain somewhere between three to ten times as many non-human cells as it does human cells. Knowing this statistic prompts us to reflect on questions of identity. A human being is nothing more than a very large community, on every level. Aren't our very ideas and language an ecosystem filled with concepts and words that came from elsewhere? These expressions we borrow from other people, these ideas that were given to us and that live on inside us, these stories of others—they all live in symbiosis with our own personal stories. The words and lives of others coexist and swim along side by side in our identities, as in a great coral reef.

The corals we see on the rocks of the Mediterranean—the yellow "pig-tooth coral" that grows on rock, and the alcyonacea

with their large soft flowers—are less vibrant than their tropical counterparts, but they tell the same riveting stories. It's
in this very sea that coral first originated, according to Greek
legend, when the hero Perseus faced off against the monster
Medusa. Medusa was one of the formidable Gorgon sisters,
who all had snakes for hair and could turn to stone anyone
who met their gaze. We don't know whether Perseus was inspired by the sardine's stratum argenteum, but he had the idea
to use a mirror on Medusa, so as not to look his adversary in
the eye. Perhaps light lost its petrifying magic when polarized
by the reflection? Whatever the case, Perseus was thus able
to behead Medusa, and as he relished his victory, a few drops
of Medusa's blood spilled onto some seaweed on the shore,
instantly petrifying it into coral. In their legend the Greek
poets unknowingly posited the idea of coral symbiosis and
links between seaweed, rocks, and a tentacular animal. Even
better, according to the poets coral was born of the monster
known as Medusa. We know today that both corals and jellyfish, also scientifically known as medusas, are animals that
belong to the cnidarian family, close cousins that bear little
resemblance to each other but possess identical anatomical
functions, and are often one and the same. Actually, most
jellyfish can at some point in their lives fasten themselves to
a rock and morph into polyps. Conversely, most polyps are
capable of setting off to live in the open water in the form of a
jellyfish (with the exception of reef-building corals, who only
live in open water when they are still larvae). Meanwhile, the
Gorgons, Medusa's two more obscure sisters, gave their name

to a genus of deepwater corals that grow without any need for sunlight.

Jellyfish repel many people from the water. They provide an excellent reason to stay and tan on the beach. But I remember, when I first discovered the magic of diving, what fascinating creatures they were to observe. Nothing could get me out of the water, even when daylight began to fade and the water turned chilly. As I stood shivering, wrapped in my towel, I was already dreaming of diving back in, to hear more stories from these sea creatures.

Alas, summer vacations inevitably came to an end.

ARE FISH GOOD
AT SCHOOL?

In which the sole is flattened.

In which anchovy eat their own eggs.

In which whales exchange song lyrics.

Childhood memories. "The square of the hypotenuse is equal to the sum of the squares of the other two sides." I had hardly taken a seat at my desk, and already the teacher's monotone words were putting me to sleep. The ring of the school bell in the playground reminded me every hour, rather viciously, that summer vacation was well and truly over. Outside, the weather was still beautiful.

Why did we have to go back to school in September? Had someone made a deliberate choice to bring us back just when summer is at its loveliest, when the sea is calm and steeped in sunlight, and the leaves of the trees display all their colors in defiance of the shortening of the days? Had they considered that this is the cruelest season during which to confine children, when the sky glimpsed through the window is still bright, calling like a siren's song, luring us out to adventure and freedom?

Head in my hand, elbow against the top of the particle-board desk, I barely listened to the tedious nomenclature of Euclidean geometry. Like a chicken in a battery cage, I had no idea why I was there. To learn things, people told me. Yet it was the teacher, the one who knew things, who was asking the students questions, and not the other way around! This teaching method struck me as quite enigmatic.

Half asleep, I extracted a "regulation-size, large-format, wide-margin sheet of graph paper" from my notebook and started doodling distractedly with my No. 2 pencil. Even this pencil, required by every official school curriculum, had been imposed with no explanation. Everyone now reading this page probably used those pencils throughout their childhood without ever being told what its number meant, and without ever learning about other varieties of pencil, such as the No. 4 or the exotic 6H's or 8B's. Beneath the graphite tip of my pencil, the dreamy outlines intertwined, creating a landscape that engaged my mind and my gaze, gradually leading me further afield from the equation-packed blackboard. The lines broke free from the checkered squares of the graph paper, gradually dissolving that cage of light-blue bars and the red margin. I bobbed along the waves of the drawing. A sardine appeared beneath my pencil tip, slowly developing and taking shape. I could already hear the sound of the surf in the scratching of graphite on paper. The classroom was fading into the blur of a coastal fog. Little by little the fish grew in size. Its image carried me off with it.

Fish don't need to go to school. They have their own ways of learning all they need to know in their fishy lives. Nonetheless, their childhoods are rather complex. When they are born, most fish—whether they are destined to one day become little sardines, or sea bream, or even giant tunas or swordfish—emerge from an egg barely one millimeter in size. They are just tiny underdeveloped larvae, shot out into the open water among the plankton. They don't know how to swim, how to feed themselves, or even how to breathe. They just drift, feeding off a yolk sac, capturing oxygen as it diffuses through their skin. They have a lot to learn.

But they're quick studies. In just a few days' time, the larvae begin to contract their muscles, then to move with more control and pursue tiny prey amid the plankton. Their gills and fins begin to develop. That's when they start to discover their stories, what fates await them. An eel larva learns that it will migrate toward shore, and starts to swim vigorously in the Gulf Stream. Salmon larvae immerse themselves in memories of their native stream, beginning to recognize its scents so that someday they can find their way back to those child-hood recollections. The larvae of coral reef fish, for their part, learn to listen. They strive to pick up the murmured sound of the distant coral-reef dwellers, and then they set off, swimming toward that singing until they reach its source. These tiny hatchlings launch into a journey that will last many long months, winding through the empty expanse of the open sea.

For fish larvae, traveling through water, especially on such lengthy treks, is quite a complex task. Water doesn't behave for such tiny creatures as it does for beings of our size. From the point of view of a small animal looking at water from close up, the diffuse movement of the water molecules far outweighs the effects of inertia and the convection of the larger currents. On a small scale water doesn't move as a cohesive unit; instead each of its molecules has its own chaotic movement. The smaller an object, the greater the impact of these chaotic movements, and the more drastically the flow of water around it will be slowed down as a result of this agitation of water molecules. A tiny creature, therefore, perceives water as a motionless, extremely viscous fluid constantly holding it back. For something the size of a fish larva, water is every bit as viscous as honey is to us. As it grows larger, the fish will free itself from that viscosity, and feel the water flow differently around it, as a lighter, more slippery fluid. It will be able to gather momentum, glide freely, and even drift along with the movement of its wake. As it continues to grow, the fish will have to relearn constantly how to swim. Relearning, rediscovering—this describes our destiny as humans as well, especially during our school years. At first we ask little children to be creative, to draw freely on a sheet of paper. Then we tell them to color inside the lines, to respect the rules. Compose orderly sentences—subject, verb, object. Meet deadlines. Then come exams in which we're again expected to be original, but within a given box, without taking risks. And when school ends, everyone must relearn in their own way, inventing their own rules or discovering their own originality.

Adapted to its planktonic life, a fish larva doesn't remotely resemble the animal it will become once fully grown. A young ocean sunfish resembles its namesake; it's surrounded by triangular rays like the sun on a child's drawing. A sardine larva is a sort of filiform eel. A swordfish larva looks like a dragon, with no swordlike bill but bearing an immense sail fin bristling on its back. This larva is very seldom observed, because it grows rapidly during its first few weeks, attaining a weight of roughly eighty pounds in a year. Flatfish larvae, such as sole and flounder, are born "normal" fish. They swim in the open water and have one eye on each side of their head. One of those eyes will migrate as the fish grows, joining the other eye on the other side, which will gradually flatten. This must mark a strange perspective shift for the young sole. It abandons the freedom of open water in order to fasten itself to the seabed, merging into the sandy bottom and covering itself with sediments that give it its color. It grows used to looking at the world from underneath, the sky its only horizon. From then on, it lives flat, in a two-dimensional world.

Like the sole, my pencil fell to the floor, hitting the linoleum with a muffled sound. At a sharp glance from my teacher, I quickly and instinctively slid the sheet of paper under my notebook, to keep him from seeing my drawings. He went on lecturing without the slightest inflection in his voice. I'd made a narrow escape, but now he knew to keep an eye on me.

METAMORPHOSIS OF THE FLATFISH

The teacher went on talking, without making eye contact, far more boring than the most silent fish in the sea. He peered down at us from a distance, but only to make note of who wasn't listening. I tried to pay attention to what he was saying, but I couldn't fight the boredom. The waves were calling out to my heart. On my graph paper were seagulls, a number of abstract arabesques, and an unfinished landscape begging to be completed. Discreetly I pulled a corner of the sheet out from under my notebook. On the alert, I started coloring with the four colors of slightly sticky ink in my pen, ready to hide it away at a moment's notice. That forced secrecy struck me as ridiculous. What other creature on earth is expected to learn while being stalked by the very person who claims to be his teacher?

Nevertheless, compared with some sea creatures, I had no reason to complain. There are fish whose own parents hunt them from the day they're born. Anchovy, unable to tell the difference between the eggs their females have just laid and the plankton they feed on, devour 28 percent of their own progeny. Pike are more patient, and wait until their young have had a chance to grow before eating them. The females of the species even devour their male partners just seconds after spawning. But as surprising as it may seem, these behaviors, given that they've persisted over the course of evolution, are in the best interests of the species' survival. After spawning, the fish are exhausted, and they need energy-cheap calories if they hope to survive. Nice fat eggs or an exhausted male are more useful and readily available, to that end, than their usual prey.

The freshwater sculpin, a stocky fish that lives at the bottom of mountain streams, has developed a clever compromise to solve this dilemma between the success of the spawn and the survival of the parents. The male guards the eggs in a sort of underwater grotto, on the ceiling of which each female separately lays a clutch of eggs. In order to protect the eggs, the male must stop eating for a full month. If his hunger becomes too great to ignore, he'll sparingly gobble a few eggs from each clutch, never consuming the entirety of any one. That way, there's always a certain number of yolk-sac fry, also known as alevins, that hatch from each clutch of eggs, and therefore from each individual female, thus assuring the genetic diversity of the new generation.

The sand tiger shark, on the other hand, has a more radical strategy. It is ovoviviparous, meaning its young hatch from an egg, but inside the mother's womb, and then continue to grow until birth. In the meantime they have no umbilical cord to nourish them, so their feeding strategy is far more aggressive. A single female sand tiger shark mates with a number of different males and so carries many dozens of shark embryos, sired by many different fathers. The first ones to hatch, stronger than the others, then proceed to devour their step-siblings while still in utero, then the unhatched eggs, and eventually even the unfertilized ones. By the time they're strong enough to face the outside world, there are never more than one or two survivors, roughly a meter in length. This intrauterine fratricidal battle results in the natural selection of only the strongest embryos, the ones with the best chance at survival.

Fortunately some fish enjoy a much gentler childhood, or at least one that more closely resembles our own. Many fish parents protect their eggs and watch over their young. Among fish, the males are the ones that typically watch over their progeny. They perform that duty with immense, truly exemplary devotion, miles ahead of their human counterparts. The male lumpfish, a perfectly round creature from the cold waters of the northern seas, whose black or red eggs are sold as substitutes for caviar, oxygenates his mate's eggs in a seaweed nest in shallow water. The male attaches himself to a nearby rock with a ventral sucker, and then keeps watch over his spawn for six to seven weeks, protecting the eggs until they hatch. Certain tilapias in Lake Tanganyika are even more devoted parents. In order to better protect their offspring, they fertilize and incubate the eggs in their mouths, then raise their young in the same place. Living in their parent's mouths, the little ones feed off the same nourishment their parents absorb. As for seahorses, the female deposits her eggs in a pouch located inside the body of the male. The seahorse father then takes care of fertilizing and carrying the eggs, until the day he gives birth to hundreds of baby seahorses that burst out of the ventral pouch like fireworks.

Relatively few fish live in family units, but it does happen in certain species, such as the clown fish. In their sea-anemone homes, clown fish form a strange family. There are the two parents, a male and a female, and their offspring, who are all born male. If the female leaves, her partner will turn into a female, and then the most mature of the young males will take over his role as the husband. Had a certain famous animated film chosen to hew to that reality, it would have affected the plot significantly.

Many fish species are hermaphroditic in this way, changing gender over the course of their lives. The underwater world is very open-minded about these "gender issues" and displays a great array of diverse behaviors. On the Mediterranean inshore reefs, wrasses, a key ingredient of bouillabaisse fish soup, are all born female. With age they turn male, decking themselves out in brighter colors and a vermilion stripe. But when the time comes for their gender transformation, certain wrasses prefer not to adopt the male livery. They instead keep the appearance of a female, even though their internal organs are turning into male organs. While the other males fight for the favors of the females, these feminine-looking males can discretely insinuate themselves into the good graces of the real females and seduce them without attracting the slightest suspicion from their masculine partners. The wrasses that promenade and swirl over the rock beds of the Mediterranean Sea are a joyous and colorful spectacle. As soon as something happens, they come sailing through the water from all directions and flip like acrobats in a circus ring. Often while diving, I'd admire them and say to myself . . .

"Let's see your workbook!" The sinister rumble catapulted me violently out of my subaquatic reverie. This time, the predator had won. "Two hours' detention."

A Wednesday afternoon spent trapped indoors in confinement, deprived of my personal freedom. What punishment could be crueler for a child than this fine paid in hours of his life? Sentenced for daydreaming and drawing little sketches instead of studying the angles of triangles, I joined two fellow students who were serving their time in a grim and empty classroom for "talking in class." For these teachers who were supposed to instruct us about life, the worst possible crimes were speaking and dreaming.

And yet communication is indispensable to learning. It's fundamental, even, to a civilization's founding and survival.

Octopuses are considered among the smartest animals on the planet, and they likely hold the record for intelligence among invertebrates. Their brains are astonishingly powerful, capable of reasoning and deduction. They are a veritable anomaly of evolution within the mollusk family, which mostly includes creatures that are rather simple-minded, at least at first glance, such as the mussel or the periwinkle. In addition to their sharp mind, octopuses possess astounding bodies. They are utterly supple, capable of worming their way through tiny holes; they can change shape and color at will; they possess eight arms that each house part of their nervous system—tools whose sharp agility would rival our most advanced robots. All these assets ought to have made octopuses our planet's dominant species, especially when you consider that, as ocean-dwelling creatures, they could construct their civilization on 71 percent of the globe's surface.

But octopuses have not succeeded in ruling the world, at least not yet. One possible explanation for their failure might be the way they transmit knowledge. An octopus acquires knowledge throughout its lifetime. It develops complex strategies, such as the ability to disguise itself as one of its predators in order to escape their attacks, or that of using empty seashells as armor, or even the habit of holding its breath while climbing onto dry land to hunt for food. Until recently people believed that these cephalopods were incapable of communicating, but now we know that's not the case. They trade survival strategies, speaking through arm gestures or changes in color, and there are even octopus towns, built atop the shells of the creatures they feed on and governed by complex social interactions. But even though they're capable of sharing knowledge among themselves, octopuses remain incapable of handing it down to subsequent generations. This is due to their method of reproduction. The start of an octopus's life is quite tragic. Once the eggs have been fertilized, the male flees in pursuit of other occupations, and the female remains in the cave where the eggs have been laid, watching over and oxygenating those little white stalactites in which the octopus embryos wriggle and squirm. But the eggs take such a long time to hatch, and the female is so devoted to their protection that, having completely neglected to eat after depositing her clutch of eggs, she starves to death before her young can hatch. She will never be able to converse with her offspring, nor to transmit her knowledge to their new generation. A young octopus is therefore obliged to rediscover everything for itself. This inability to educate their young may have cost the octopuses their potential conquest of dry land, and with it the creation of cities, cathedrals, 4G satellites, subways at rush hour, social media arguments, tax paperwork, and all the other delights of modern civilization. Perhaps they're better off, but it's regrettable all the same. Octopuses would surely have known to equip their cathedrals with fire extinguishers and their subway systems with Wi-Fi.

OCTOPUS

Contrary to octopuses, humpback whales raise their young for a considerable amount of time, and they communicate with them constantly. This method of calf-rearing has led to the creation of what could reasonably be described as a humpback-whale culture. Groups of whales develop cultural traits, behaviors distinctive to them that they preserve over time and pass down through a form of social apprenticeship. For example, whale song is handed down year after year among whales in a given group. Each individual adds or modifies certain verses, which are then shared with the others. The songs evolve continually over the years, in much the same way that musical styles or languages evolve. Themes appear, other themes disappear, and some transform.

In the 1980s shoals of herrings deserted the Gulf of Maine

due to industrial overfishing. All over the world humpback whales herd herrings by blowing strings of bubbles around them, so that they can swallow whole schools of them in a single gulp. When these fish disappeared from the region, the whales off the coast of Maine were forced to turn to another prey, sand eels, whose schools were harder to herd, as they stayed dispersed near the water surface. So the humpbacks invented a new technique, generating bubbles by slapping the surface with their tail flukes, thereby forcing the sand eels to dive. These whales handed down this new sand-eel fishing method from one generation to the next. A "naïve" whale from another region would not naturally know how to fish for sand eels, but if it happened to meet whales from Maine who revealed their own techniques, it would be able to do it. This transmission of acquired rather than innate knowledge, by means of education and not instinct, demonstrates the transmission of culture among whales.

But I was not a whale calf, and clearly I wasn't spending my Wednesday afternoon confined in a classroom learning anything or improving myself. All I feared was what fate had in store for me, the punishment imagined by our jailer.

Ordinarily we would spend those two hours writing out in longhand the school code of conduct, or even worse, specific articles of this code, which would only steel our determination to invent a thousand ways to break them, even as our aching fists transcribed them. But our monitor, in a good mood that day, showed us mercy. Without once taking his eyes off his newspaper, he instructed us to spend our two hours of detention composing an essay. Subject: "What I Did on My Summer Vacation."

I set to work, at first half-heartedly. It struck me as the height of cruelty to be forced to remember my vacation at the very moment when I was most lamenting its end. But the words lined up in rows on the paper, and little by little they coaxed me along with them, just as the lines of my drawing had done before. I described the inlet with its luminous waters; the shallow-water grass beds, purplish green in color, undulating beneath the reflection of the sky; the gray mullets frolicking on the surface. In simple inked letters, I set the scene with schools of translucent silversides, the blue back of the sardine, the pointy arms of the starfish, the sea urchins and their loud cracklings. I soared free of my fetters and swam through the universe I was describing. All of the atrocious grammar rules that had made me suffer so, now melted into the harmonious underwater seascapes. The stylistic devices that our French teacher had beaten into our heads now came to life in the depths. The octopus camouflaged among the rocks became a metaphor for the seabed; the conger eel lurked in its hole like a euphemism, revealing only its snout. The dreamfish lined up one after the other like an anaphora; the tiny ocellated wrasse swelled to hyperbolic size to seem bigger. I felt an entirely natural pride in letting myself be lulled by that poetry, and teasing it out of the waves and setting it down on paper. I was happy to be able to share it.

But the only sharing I did was with the paper. As soon as we handed in our copies, our monitor stacked them distractedly in a pile and, without even a glance, crumpled them up and tossed them in the trash.

Just ten months to go, I told myself as I left the room. Luckily the cold weather would soon return, and with it the Christmas vacation. I would be able to listen to the stories of the sea again, but in a very different way. Because it would soon be seafood season.

COCKLES AND
MUSSELS

In which, even if you don't love oysters, you'll have things to say about them at your next seafood dinner.

In which a whelk reunites the Jewish people after a two-thousand-year quest.

In which distant galaxies glitter in the black eyes of prawns.

Seafood has one thing in common with cilantro, strong cheeses, and licorice-flavored tea: it tends to divide people. Although no one is really born with a love of oysters, there is a clear distinction, past a certain age, between those who sincerely love them, those who pretend to love them but mask the taste with vinegar, and those who openly despise them.

Among the true oyster-lovers, only a select few actually know how to properly open them without falling back on often dangerous tricks touted in countless online tutorials.

But whether or not you appreciate the taste, opening an oyster is like opening a book. You are presented with a drop of seawater enclosed in the mille-feuille of the shell, a pearly treasure safe under a rocky crust that resists and refuses to open. An oyster is full of underwater rumors and oceanic stories that, sealed inside, are not often shared.

While the gourmet pries open these shells and slurps down their contents, let's wait for the oyster to calmly open up and release a few of its secrets.

Even on the outside, the oyster shell is a material unlike any other. The mother-of-pearl, or nacre, from which it's made is a biomineral, a mineral produced by a living being. This alliance between the animal and mineral kingdoms gives it extraordinary properties. Mother-of-pearl is composed of 99 percent calcium carbonate, also known as chalk. But the oyster knows the secret art of transforming soft and crumbly chalk into tough and precious mother-of-pearl.

The remaining 1 percent of ingredients contained in mother-of-pearl is a secret recipe for a protein-based cement, which the oyster uses to transform the chalk. The technique is still a mystery, but we do know that the oyster adds certain mineral salts to the chalk, in order to turn it into tiny tablets of limestone crystals called aragonite, which measure about a dozen microns across. The oyster then glues these crystals together in an unknown process involving a protein named conchiolin. Glued together in this way, the crystals form a material three thousand times stronger than aragonite alone, and of course much stronger than pure chalk. Mother-of-pearl by itself has no color; its material is unpigmented. But when sunlight falls on it, it reflects off each tiny aragonite tablet. These tablets are so small and so regularly spaced that the reflected beams interfere with each other, breaking down the sunlight into clearly distinct colors, which in some shells produce beautiful rainbow-tinted iridescences. Opticians call these "structural colors," meaning that the colorless material of mother-of-pearl, through its shape and structure, breaks down the light into colors without itself possessing any trace of pigment.

The oyster tirelessly produces mother-of-pearl not only in order to grow, but also to protect itself. If a grain of sand enters an oyster's shell, it's like a pebble in our own shoes—annoying, distracting, and before long, painful. That's why the oyster starts turning the grain of sand over and over, while covering it with mother-of-pearl, in hopes of ejecting it or at least making it smoother. Little by little, mother-of-pearl is deposited on the grain of sand, rounding it into a pearl. All oysters produce pearls. Although it's rare to find pearls in commercially sold oysters, it's not entirely impossible, and when it comes to pearls, it doesn't hurt to believe in miracles. Oysters know this well. To help you keep the faith, even in front of your next oyster platter, let's listen to an oyster who opens up slightly and tells us the true modern tale of the world's largest pearl.

This story unfolds somewhere in the distant, crystal-clear waters of the Philippines. In the coral reefs of those latitudes live the world's largest mollusks, measuring more than a yard across. This is the *bénitier,* French for giant clam, but also for "holy water stoup," the large vase used in Catholic churches to hold holy water. These mollusks take their French name from the role they played during the Renaissance, when explorers brought them back to Europe as vessels for holy water. Some still serve this purpose today in old European cathedrals.

One day a grain of sand wedged its way into the shell of a giant clam, somewhere in the bed of a coral reef in the Palawan area of the Philippines. All the giant clam's efforts to expel the grain of sand were fruitless. The mollusk had no choice but to produce a pearl around it, and that pearl grew and grew until it took up nearly all the room inside the shell. Things took an unexpected turn when a great storm hammered down one evening in the early 2000s. A local fisherman out at sea was unable to return to shore because of the breakers surging over the barrier reef. Making up his mind to spend the night at sea, he dropped anchor in the shallows. When the storm died down the following morning, and it came time to weigh anchor, he realized the anchor's fluke had gotten stuck under something and dove down to yank it free. To his surprise he discovered that the anchor was stuck inside an enormous clam containing an immense, coiled, pearly mass with strange circumvolutions.

This fisherman was very poor and very superstitious. He had no idea what the pearl was, but he believed it was a magical object. Back home he hid it under his bed. Every morning before he set out to fish, the fisherman would touch the pearl under his bed, convinced that it would bring him luck. Ten years went by. Sometimes the fishing was good, and sometimes the fishing was bad. With a sailor's typical confidence in the supernatural, the fisherman remained convinced that the magical object was watching over him.

After ten years the Filipino fisherman decided to move house. His aunt, who worked at a tourist museum in the city, came over to help him load boxes. When she laid eyes on the pearl, she was quite amazed, and recommended that her nephew have it appraised by an expert.

The story doesn't tell us whether money brings happiness. But the Filipino fisherman found himself the owner of the world's largest pearl, weighing in at seventy-five pounds, for an estimated value of more than twenty million dollars. Perhaps he was right to believe it was magical.

For many centuries pearls were very rare objects in France. It was therefore the artificial pearl industry that supplied the finery for the most spectacular courts in Europe. These artificial pearls did not come from oysters, or even from the sea, but their story nevertheless originates with a fish—an unassuming freshwater fish, a kind of shiner found in the waters of the Seine in Paris and the Rhône in Lyon, the common bleak. The invention of the method for producing artificial pearls is itself a story worth its weight in mother-of-pearl.

The year was 1686, in the Paris region. A *patenôtrier,* or rosary bead maker, named Maître Jacquin was bemoaning the very trade that had made him his fortune, fake pearls jewelry. Like all his competitors of that time, Jacquin gave the glass pearls he manufactured a nacreous appearance by filling hollow glass spheres with a mixture of mercury and lead that was terrible for the health of his customers. He knew it only too well, and so did his customers, but nonetheless those fake pearls were flying off the shelves, and people were paying their weight in gold for them. With each passing day Jacquin sank deeper into despair.

And yet he ought to have been ecstatic. His son was about to marry the ravishing Ursule, daughter of a neighboring apothecary. But the moment Maître Jacquin had been dreading had arrived. Ursule had just entreated Maître Jacquin to make one of his poisonous pearl jewels for her wedding day.

Unable to bring himself to fulfill the request, he spent long hours pondering a solution. As he wandered one day along the banks of the Seine, the enameled gleam of a school of common bleak caught his eye.

The rosary bead maker had no way of knowing that the scales of those fish possessed the same microscopic tablet structure as mother-of-pearl, in arrays of cells known as iridophores, and hence gave off the exact same iridescence as pearls. But he had a hunch that the brilliant glimmer would have a similar effect. With the assistance of his future brother-in-law, the apothecary, he developed a procedure using ammonia to preserve the tiny scales of the common bleak and inject them into the fine bubbles of wax-filled glass. He dubbed this process *essence d'Orient*, Eastern spirit, and soon all the courts of Europe were lining up to purchase these iridescent but harmless faux pearls.

Since it took twenty thousand fish to produce five hundred grams of *essence d'Orient*, the common bleak fishing industry provided a living for entire villages along the banks of the Seine, the Saône, and the Rhône for more than two hundred years. The wheels of many watermills, originally built to crush the scales of the common bleak, still turn in many French villages.

Beneath the oysters on most seafood platters, there is often a layer of discreetly hidden tiny sea snails, or periwinkles. Why serve periwinkles on a seafood platter? No one ever eats them, as extracting the meat from a periwinkle requires the dexterity of a chimpanzee surgeon, and yet with every platter, the chefs insist on serving an ever more immense pile of these gastropods, occasionally without even providing the indispensable side of mayonnaise or the crucial picks.

They often add whelks to the mix. And what could be more silent and less interesting than a whelk?

And yet the whelk, inspired by the oyster, now opens its cap and begins to tell a story, the story of its Mediterranean cousin, a whelk that incited a millennia-long quest spanning the four corners of the world. A quest as old as the Bible itself.

So it is written in the Old Testament. The Almighty commanded Moses "Speak unto the children of Israel and bid them that they make them fringes in the borders of their garments throughout their generations, and that they put upon the fringe of the borders a thread of Tekhelet." This was a sacred color. Tekhelet was both "black as midnight" and "blue as the sapphire of the Tablets of the Law." It was at once "blue as the sky around the sun" and green. So it is written. Tekhelet was sacred because it came from the hillazon, and the hillazon was a mollusk that "resembled the sea," and the sea resembled the sky. So it is written.

For centuries the Hebrew people produced the tekhelet dye from the mollusk hillazon and used it to adorn the fringes of their garments. Extracting this heavenly gift from out of the sea, then tinting the woolen fringes with this divine color was an ancestral ritual.

But the Hebrews weren't the only ones to produce colors from mollusks. The Greeks and Romans had their own pigment from the waters, Tyrian purple. It was not a divine gift. In their own tales, it was Hercules who had discovered the Tyrian purple, or rather his dog, whose lips turned purple from chomping shellfish on the beach. This purple lacked the brilliance of tekhelet. Actually a pinkish violet, almost burgundy, it wasn't the color of God, but the color of glory, emperors, and persons of note.

Since producing a single gram of Tyrian purple required extracting by hand the contents of twelve thousand murex shellfish, the dye was costlier than gold. Its trade solidified the glory of the Phoenician city of Tyre and brought in millions of sesterces, attracting unbridled greed. Caesar himself deemed this a good way to replenish the coffers of Rome, and therefore he decreed that all dyes made from seashells were henceforth an imperial monopoly.

Tekhelet was not exempt from this decree, and in order to continue honoring their divine commitment, the Hebrew dyers defied the law. For nearly two centuries tekhelet became a clandestine dye. It was worn discreetly in the streets of Jerusalem; everyone knew where it came from, but no one breathed a word. The Romans turned a blind eye, since all they cared about was the pinkish hue of Tyrian purple. But the mad emperor Nero wanted to be the only one to wear purple of any kind. He issued a law reserving for his own personal use any color that came from the sea, enforcing it with the strictest violence throughout the empire. The Hebrews could do nothing then but resign themselves in defeat and despair to the new prohibition.

The existence of tekhelet remained documented in the Bible, but as generations passed, the art of its production was lost. The ḥillazon mollusks lived happily on the reefs of the Mediterranean Sea. Soon no living human even knew what they looked like.

Over the course of many civilizations, the Jewish people scattered across the globe, far from the shores where the hillazon carried its polychromatic secret. And yet the rabbis kept the memory of that lost color. They had never even seen the color's gleam, but they knew it was their duty as believers to do everything in their power to find it again. Alas, the words of the Bible did nothing to facilitate their task. The color was described variously as black, blue, and green. As for the hillazon, it was written only that it had a shell and "resembled the sea."

In medieval Spain the eminent rabbi Moses Maimonides decided that tekhelet was probably a light blue, and the Sephardic Jews in Northern Africa began to decorate their prayer shawls, or tallitot, with blue fringes. In the same period in Burgundy, the respected rabbi Rashi of Troyes reckoned that tekhelet must have been black, so from then on the Ashkenazi Jews of Europe wore their tallitot with black fringes.

But for those who wished to follow the biblical example to the letter, all efforts proved useless to find a mollusk that resembled the sea and produced a pigment that was simultaneously black, green, and blue. The violet sea snail, *Janthina janthina,* a shellfish as blue as the sea, secretes a light-blue colorant to fend off predators, and it was a candidate, but that color was only pure blue, without the slightest hint of green or black. In the nineteenth century, the Radzyner Rebbe had another idea: the hillazon might simply be the cuttlefish. That cephalopod is indeed capable of changing color, like the sea, and of camouflaging itself by taking on the appearance of the seabed. It possesses a shell of sorts called the cuttlebone, and it emits a black ink. They just had to turn it blue. The Radzyner Rebbe developed a chemical treatment that allowed him to obtain indigo from cuttlefish ink.

The hope that the hillazon had been found lasted until advancements in chemistry made it clear that the blue pigment thus obtained was formed, not from cuttlefish ink, but from the carbon atoms left behind when it was burned. In other words, any carbonized organic matter would have provided a blue dye through the Radzyner Rebbe's procedure. The cuttlefish wasn't the hillazon after all.

Some Jews wound up believing that God had intentionally withdrawn the hillazon from man's dominion, and that only the Messiah would bring back its secret.

But in Lebanon in the 1970s, archaeologists discovered the ruins of giant warehouses full of murex shells, which offered researchers a new lead. What if this murex, which the Romans cultivated for their own imperial purple, had a secret other identity?

The murex was a large striped whelk. It certainly hadn't resembled the sea when the rabbis had seen its shell on museum shelves. But the living mollusk in the water covered itself with seaweed and various concretions, taking on the rocky and mossy appearance of the seabed.

It was likely a fisherman who, shattering a murex one bright sunny day, uncovered the secret of tekhelet when he saw its juices turn first black, then green, and finally blue. It would take another two thousand years for scientific proof, but in the 1980s the chemist Otto Elsner provided a demonstration. Murex pigments changed color under the effect of sunlight's ultraviolet rays, turning blue, black, or green. The lost hillazon had finally been found.

The prayer shawls of Jews all over the world were finally reunited with their history. All of this thanks to a whelk, a whelk whose flavor no rabbi would ever know, because eating shellfish is prohibited by the Old Testament.

Alongside the oysters that are impossible to open and the whelks we all ignore, every seafood platter also contains prawns.

A prawn is a harmless, unremarkable creature, and yet have you ever noticed how deeply strange it is? As you peel one, do you ever stop to imagine for a moment what it must be like to be in the skin of a prawn?

A prawn carries its skeleton on the outside of its body. And it changes its skeleton practically every month, meaning that it's completely soft and defenseless for several days, just long enough for the new skeleton to grow out. A pretty bizarre existence from the jump.

Prawns are also great talkers. They communicate primarily by touching antennae, which they also use to taste and hear. For prawns hearing, taste, and speech all blend together.

But have you ever noticed their disconcerting, completely black eyes?

What makes a prawn's eye so radically different from our own is that, instead of being transparent and focusing light through a lens, it is completely opaque. Even stranger is the way it absorbs light through tiny alveoli whose interiors are lined with mirrors. These little wells condense light at their centers, where the optic nerves are located, giving the prawn incredibly strong eyesight. It can see 180 degrees around itself, even in very dim waters.

At night, in the depths of the sea, plankton gleam like galaxies. Prawns can admire this spectacle in their dreams, watching jellyfish float overhead like shooting stars.

The sea resembles the sky, said the whelk in the Bible. One day human beings had the idea of imitating the prawn in order to gaze at the sky. That's how NASA's telescopes, modeled on the prawn's eye, observe the X-rays emitted by galaxies on the distant edges of the universe.

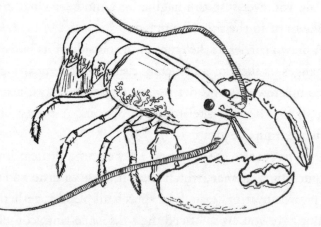

LOBSTER

But the real star of the seafood platter isn't the prawn. It's the lobster. Of course the costliest and most refined delicacy should be the one that's impossible to enjoy without splattering the walls from floor to ceiling, so necessitating the use of a hammer, some kind of screwdriver, and several bandages. Even, in some restaurants, a ridiculous bib to avoid getting stains on your suit.

But the lobster thoroughly deserves all this honor.

It hasn't always reaped such glory. Because it closely resembles a monstrous insect, it was long reviled by gourmets. State prisoners in New Jersey two centuries ago were overjoyed when the jail's administrators promised to stop serving them lobster more than four times a week. The ninth amendment of the Constitution specifically prohibited such a "cruel and unusual punishment."

Admittedly this animal is even stranger than its smaller cousin, the prawn. It uses its hairy claws to taste the water. It urinates through antennae located just below its eyes, and that is also how it communicates, except when it prefers to emit loud roars. Lobsters live in underwater caves alongside conger eels, which offer them the scraps of their own meals, and wait to devour the lobsters the moment they lose their shells in the molting process. A lobster will grow indefinitely for the entirety of its natural life, and if it loses a claw or a leg or an eye those parts will promptly regrow. This crustacean can even amputate one of its own limbs to escape imminent danger, knowing it will regenerate the next time it molts. During the molt the lobster emerges from its old shell and devours it; the old shell provides the calcium needed to build a new one.

The lobster's liver, which also serves as its kidney and pancreas, is the green part in its head, appreciated by only a select few gourmets. The orange part is made of eggs, which will hatch into adorable baby lobsters, as round as crabs.

Melted butter, on the other hand, does not occur naturally within them.

The lobster is too bizarre to tell a story that everyone would understand. It lurks quietly, dreaming from time to time. Perhaps it thinks of its old friend, Gérard de Nerval. This Romantic poet, the author of mysterious and esoteric verse, fell victim to fits of madness at the end of his life. So he adopted a lobster as a pet, and would walk it on a leash of blue ribbon through the streets and cafés of Paris. When passersby gawked in astonishment, he would reply insolently that he much preferred this animal to a dog. "I have a fondness for lobsters," the poet would say. "They are peaceful, serious creatures. They know the secrets of the sea . . . and they don't bark."

Of all the forms of seafood, mussels, though certainly less noble than the lobster, are probably the food people most agree on. *Moules marinières* with french fries. Say the words, and you can already smell the salt spray, evocative of summer vacations and conviviality. These shellfish have strange stories to tell. Both the female mussels, recognizable by their orange flesh, and the males, more of a creamy yellow, are astonishing creatures, capable of filtering sixty-five liters of water a day in order to feed on plankton. But mussels are known especially for their ability to fasten themselves to their rocks with an incredibly tenacious grip. They do so with the help of their byssus, a bundle of filaments with extraordinary adhesive properties. It can even attach to Teflon. It was while adhering to a rock that the world's largest mussel, which lives on the Mediterranean coast, gave rise to a surprising story.

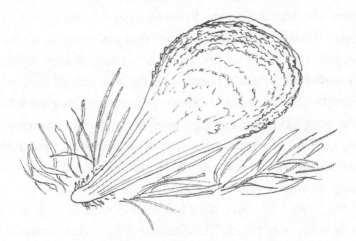

NOBLE PEN SHELL

Without fail, each time I dive in the Mediterranean, I greet a number of respectable creatures as though they were old friends. The noble pen shell is one of such longtime companions. Even now it strikes me as unthinkable not to stop during a dive to gaze at this venerable shellfish with its hues of translucent amber, peppered by the passage of time with red algae and gold-colored wormholes.

The noble pen shell is the world's largest mollusk. Its

pinkish shell, more than a yard high, rears up vertically above the meadows of posidonia seagrass, and this particular ovular shape is responsible for its nickname, "the big ham."

I never would have imagined, the first time I saw one, that this mute, unassuming shellfish was at the root of the legend of the Golden Fleece!

A staple of Greek mythology, this legend tells of the prince Jason, who had to find the fleece of a golden ram to win his father's throne. His quest unleashed many adventures, including, per the folklore of the time, the slaying of dragons, the taming of fire-breathing oxen, and the mincing of innocent children. The connection between a Greek hero on a quest for a shearling jacket and a giant mussel of the deep isn't immediately obvious. If I add that it was on the Himalayan foothills that this shellfish gave rise to the legend of the Golden Fleece, things only get more complicated.

The story starts five hundred years before the Christian era, in a region of Central Asia known as Bactriana, nowadays occupied by former Soviet republics with names ending in "-stan," very useful if you're playing a parlor game that involves naming international capitals.

Try to picture the frigid steppe, the wind cutting and gritty with sand. Camels were covered with frost and exhaled billowing plumes of mist as they trudged beneath their wicker baskets crammed full of market goods. Wolves were always at their heels, staring down the caravan, while the threat of highwaymen loomed like the shadows of vultures. What a relief when the merchants finally spotted, in the distance, the fortifications of the caravanserai!

After weeks in the desert, they finally reached shelter and fresh bread. These merchants came from Antioch, a Greek port city on the Mediterranean. They were laden with bales of a fine golden fabric. If they didn't meet any obstacles along the way, their strange fabric would arrive in a few months in the city of Xi'an, in the faraway Empire of Seres, which at that time was not yet named China.

At each stop along the way, the merchants would meet other caravans, and their trade went well. Some were transporting ivory from Carthage, amber from the Baltic Sea, or spices. Others, from the distant Empire of Seres, were traveling the same route in the opposite direction. These merchants expressed a special interest in the fabrics of the Greek traders from Antioch. Their own yaks were piled high with silk.

Under arches smoky with incense, the curious Sere merchants invited the Greeks to compare their fabrics. They proudly explained that their rolls of silk were made from the cocoons woven by caterpillars before they turned into moths. Spinning a fabric out of those cocoons demanded great effort, hence the price. Then the Sere merchants insisted on seeing the merchandise the Greeks had brought from Antioch.

So, feigning nonchalance, one of the merchants opened a bale and unrolled a bundle of fabric. And as always, a long, astonished silence followed. Before their disbelieving eyes the fabric glittered like gold. It was as soft as velvet and both supple and strong. The crowd went wild. Everyone rushed to touch it, glimpse its gleam, and above all to hear its story. And with sweeping gestures, the merchant went on to tell the story of the golden fabric, sometimes stopping for a moment to let his translators catch their breath.

In vast meadows under a blue sea lived giant shellfish, firmly anchored to the rocks. These shellfish clung to the rock by means of a byssus, fine but strong filaments that were capable of resisting both raging tempests and octopuses' fiercest efforts to pry them loose. Every year, divers went down to snip those fibers. It took hundreds of shellfish to produce a few strands. Then a secret treatment involving cow urine gave the threads a golden color, and a lengthy spinning and weaving process resulted in those precious bolts of "sea silk."

Although the Sere merchants were fascinated by the quality of the fabric, the story behind it hadn't convinced them. They found it far too fanciful. A silk made of shellfish— unthinkable. The Seres therefore put forth other explanations. That golden silk was much more likely woven from the hair of the mermaids whose eyes wept pearls, as some legends claimed, or perhaps it came from a ram with webbed feet who, rising from the sea, rubbed itself against the rocks and tore out some of its (no doubt golden) wool. Chinese archives contain many a merchant's ledger books from that period, and in them traders repeatedly express their disbelief concerning the alleged origins of this "sea silk."

The Greeks therefore had to adapt their story in order to persuade their Sere customers. With each passing caravan, they abandoned the story of the mollusk and opted for that of the ram, which sold better. The byssus of the shellfish became the Wool of the Sea Ram. Soon this golden wool was being sold from one end of Eurasia to the other, without anyone knowing where it really came from. The Golden Fleece had been born.

Since then, sea silk garnered mentions in the Bible and on the Rosetta Stone; it decorated popes and emperors, and the legend of the Golden Fleece inspired generations of artists.

The next time I saw a noble pen shell after learning this story, the silent shellfish seemed to flash me an ironic smile. A simple mollusk, clinging to its rock, had fostered a legend. A creature without eyes, without a voice, whose life consisted of nothing more than filtering plankton, had become the creator of a foundational myth that was the talk of the world, echoing all the way to China.

My old shellfish friend certainly knew how to conceal its story behind its staid appearance!

But I was also moved for a sadder reason. The shellfish had perhaps cracked open to tell me the tragic end to its story. It was opening up one last time because it could no longer close its shell. A parasite, proliferating due to rising water temperatures, kept it from snapping shut, leaving the creature at the mercy of any passing octopus, inevitably eager to slide into its mother-of-pearl shell for a seafood banquet.

In spite of the efforts of its ally and symbiont, the little pea crab, which pinches the noble pen shell's gills to warn it to shut whenever an octopus is around, I knew all too well that this gaping shellfish would not live much longer.

In just one year, more than 90 percent of the noble pen shells vanished from the French Mediterranean coasts. Their last hope now is for the few surviving specimens to be taken to isolated aquariums on land to protect them from the parasite.

As I gaze at the empty noble pen shell I have at home, one of the last in the region, found empty after so many were wiped out, I can only hope that this mollusk with such an incredible imagination will somehow bounce back so it can continue to write its tale.

When these marine creatures tell their stories, they sometimes slip in hidden cries for help. It's up to us to decipher these distress signals.

DAILY SPECIALS

In which the depths lose their colors in Styrofoam packaging.

In which a cod steak dethrones Christopher Columbus.

In which we hem and haw.

Picture this: a restaurant by the French Riviera, a restaurant like all others, with sets of orange table placemats and Ricard umbrellas and a view of the port. The waiter passes out the menus. I find dining at a restaurant to be the ultimate pleasure, except for the horrible process of choosing what to order! Before deciding on a whim to just have the same thing as their neighbor, many ditherers, myself included, hem and haw for hours in a bout of philosophical torture. We reread once more the daily specials listed on the chalkboard, afraid of winding up like the infamous Buridan's ass, who died of indecision, and intimidated by the waiter who is clearly beginning to lose patience with us.

Should we choose a two-course meal with an appetizer and a main, or a main and a dessert? Since we're by the sea, might as well order some fish.

On the chalkboard is the day's special, fillet of ocean perch with saffron rice.

At 1,800 feet under the Scottish sea, the human eye sees nothing but black. Fortunately the rockfish has gigantic yellow eyes, which can spot everything that moves in the darkness. Its skin is a vermilion red. Since that color disappears completely in deep water, the fish appears muted and virtually invisible. Seeing without being seen neatly sums up this fish's life, spent at the bottom of the reefs, always on the lookout.

And this fish has seen quite a lot over the course of its life, because it lives to a venerable old age, sometimes more than a hundred years. The abyss keeps it young.

In cold water reefs the seabed is carpeted with fans and lace of black coral growing in the darkness. Soft white inflorescences of Lophelia, translucent gorgonian corals, and pale sea anemones form a ghostly landscape, where only the occasional luminescent fish passes by like a will-o'-the-wisp.

At these depths there is no sunlight, and therefore no plants. The seascape consists entirely of animals that, deprived of the nutritional bounty of vegetation, grow very slowly. The corals cannot cultivate the tiny zooxanthellae algae on which they normally feed, so they must wait patiently for the currents to bring them other microscopic prey. Their branches grow no more than one millimeter per year. The fish that inhabit their reefs also take life at their own pace. The rockfish doesn't reach full maturity for twenty years.

During its prolonged childhood the rockfish has learned to escape numerous threats, however slowly. It has figured out how to foil the tricks of the anglerfish. Brown and blistery like a flattened toad, the anglerfish is a master of camouflage, indistinguishable at first glance from the ocean floor. The eye of the young rockfish is merely intrigued by a phosphorescent plume shimmering at the bottom of the sea, like those feather toys that drive cats crazy. A reckless rockfish won't see the anglerfish itself, lurking just beneath its shimmering lure, wiggling it by means of a long filament sprouting from its head. The rockfish ventures closer, and the anglerfish need only pop open its gaping maw to suck in its prey in a sudden surge of water. Some 42,000 years ago, the Indigenous people of East Timor invented lure fishing. But fossils dating from the Cretaceous period show that, in the dark of the deep seas, anglerfish were already doing it 130 million years ago.

The rockfish will have also had frequent encounters with immense schools of spiny dogfish. These small sharks, roughly a meter in length, have beautiful green cat eyes and coppery flanks spangled with pearly dots. They swim with an elegant idleness, especially the pregnant females that carry their young for over two years. This is the longest gestation period in the animal kingdom, even longer than that of elephants. When they emerge from their mother's uterus, the young pups already look like fully formed adults. When the rockfish was young, forty years ago, it regularly encountered vast shoals of several thousand spiny dogfish. Nowadays it sees hardly any.

Sometimes, in the shallow parts of the reefs, at around two hundred meters below the surface, the rockfish might even run into a strange creature: the guillemot, a bird that moves underwater by flapping its wings, descending into the deep ocean to peck at marine worms. Despite its resemblance to a lazy penguin, this black-and-white bird is an expert navigator. It spends all year out in the open seas, free-diving to record depths, and returns to land only once per year to nest high atop the cliffs. The first leap it takes in its life, when it throws itself into the void to reach the sea far below, obliterates its fear once and for all.

But suddenly a terrible screeching rings out in the calm of the depths. The metal panels of a trawl net rake through the reef at stunning speed, so fast that nothing can escape, sweeping everything in its path into a vast pocket and leaving nothing behind but a trail of sludge. It would take millennia of painstaking work by the coral polyps to reconstruct the reef, centuries of slow growth by the rockfish to repopulate it. But the trawl net returns relentlessly, to hollow out the same scars ten times per year.

On the deck of the ship, swaying in the swell, the nylon net vomits a flood of distressed creatures, which are immediately sorted. The broken corals, the stray guillemots, and a huge teeming mass with no commercial value are all sloughed off into the churn of the boat's wake. Gloved hands pick through what remains. As soon as they're skinned and cleaned, they're frozen. The rockfish becomes a white slab, sold under the misleading label of "ocean perch," a name more familiar to customers and therefore easier to sell. The spiny dogfish, too, is rebranded. Stripped of its head and tail, the once proud shark is now nothing more than a pink cylinder sold under the name of "rock salmon," "cape shark," or "flake." It will be served as "fish sticks" in school cafeterias to grimacing children, transmitting its record mercury levels to those innocent predators. All that will remain of the anglerfish, its appearance too frightening for the market stalls, will be its tail, renamed "monkfish." The scarlet livery of the rockfish, the frenzied shoals of spiny dogfish, the cunning ruses of the anglerfish—in the sweep of a trawl net, it's all reduced to a heap of standardized pink flesh. Human creative genius, hard at work.

No, not the daily special. Let's take a look at the menu instead. Cod with *sauce vierge*. That seems less risky than some unfamiliar fish. Codfish is a safe bet. The guarantee of white, boneless flesh. A culinary comfort zone.

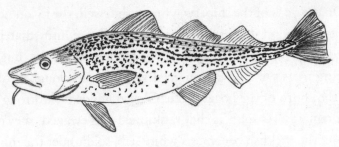

CODFISH

Some ten thousand miles to the east, in China, female factory workers stab hypodermic needles into an unending stream of fish fillets rolling by on a conveyor belt. They have no idea what they're injecting into the fillets, because it's an industrial secret—a cocktail of phosphates. They also have no idea where the fish come from—the distant North Atlantic—nor where they're off to next: the stalls of European and North American markets.

The cosmetic treatment of a phosphate injection makes the codfish appear so pearly white and appetizing, and the labor costs involved are so cheap, that the long-haul roundtrip flight is justified. Meanwhile the carbon footprint involved will only hasten the melting of the pack ice beneath which the cod once lived, protected by it for so long.

The cod has always traveled alongside civilization itself. In our freezer, where it reunites with its native ice, the final act of its story plays out, a story of voyages as old as Europe. The story of a fish that discovered America and unleashed wars and revolutions.

Cod live in cold seas around the polar circle. In order to

withstand the arctic winter and six months of endless night and frost, they form huge shoals that gorge themselves on mollusks and crustaceans all summer long, giving them their fat and protein-rich flesh.

The Vikings understood this, and figured out how to catch and preserve cod as early as the year 1000. Flattened and dried, the cod takes the name of stockfish. By now it has lost some 80 percent of its weight, but it maintains its caloric value and will keep for more than three years. For Erik the Red's longship journeys, it was the ideal snack. The Vikings relied on stockfish for the nourishment and energy required to pillage the entire European shoreline, and it was while they were ransacking the Bay of Biscay that they taught the Basques how to transform cod into stockfish.

This food brought great joy to the Europeans who, devoutly Christian at the time, ate fish every Friday and throughout the forty days of Lent, but they had no iceboxes, freezers, or meal delivery services. By the end of the Middle Ages, stockfish was exported all over Europe and became the single most widely consumed fish, leading to the development of the *aïoli provençal,* the *morue à l'auvergnate,* and a thousand other stockfish recipes developed in inland or southern regions where no living cod had ever flapped its fins, even during the Ice Age.

The lucrative stockfish trade drove the Basque sailors out west in search of the immense shoals of fish described in Viking legends and said to abound in the waters off a mythical land called Vinland, located in the unknown far western reaches of the Atlantic. And so it was that around 1390 the pursuit of cod led the Basques to Canada—specifically, to Newfoundland and Nova Scotia. They had discovered a new continent, but above all an incredible place to catch cod, a secret they kept for themselves; only a handful of maps from the period even mention it. A good fishing spot is never shared. A hundred years later, when Christopher Columbus set sail with his three caravels, their holds were already filled with dried stockfish caught by Basque fishermen in America.

During the Age of Discovery stockfish appeared on the daily menu of mariners and thus spread to each new colony. In the Antilles, in Western Africa, they're still wild about it today, served in *accras* fritters or *thieboudienne* stew. The Portuguese conquistadors, who nicknamed it "the faithful friend," exported it to Brazil and Cape Verde.

The stockfish trade went global. To gain control of the "fishing harbors" in Quebec and Newfoundland, where the precious fish yields were brought ashore, the capitals of Europe engaged in four centuries of armed conflict. This was the era of the *terre-neuvas,* the Newfoundlanders, sailors who set off aboard big three-masters to the far reaches of the polar north to fill their holds with fillets of white gold, "guaranteed boneless."

The earliest American colonies made a fortune off cod. The main export of Massachusetts, it became one of its official symbols. A "holy" wooden cod was hung over the Boston House of Representatives. Meanwhile in France the consumption of stockfish required so much salt that the king imposed a salt tax, prompting a series of revolts in 1789. You may have heard of it.

Until the twentieth century the abundance of cod seemed an inexhaustible manna, something that only the vastness of the sea was capable of offering up to mankind. Fishing techniques had remained largely unchanged since Viking times— baited handlines cast from dories, small boats with large sails. The invention of engines and freezers was about to kill the goose that laid the golden egg, or perhaps we should say the golden cod roe. Instead of fishing lines, selective and respectful of the seabed, fishing boats started raking the codfish habitat with trawl nets, in order to sweep up entire schools at a time. To remain competitive, fishermen had to catch more and more fish, which meant targeting the juvenile cod as well as the adults, and saturating the market, even if that meant eating only a portion of the haul and throwing the rest away. Waste soon became synonymous with profit. At the pace of two million metric tons caught each year, the miraculous abundance that had fed mankind for six centuries was demolished in a decade.

The shoals of cod would never return. Their habitats were colonized by lobsters, their former prey. Lobsters took advantage of the new circumstances to take their revenge by devouring cod eggs, preventing their return in spite of all the measures protecting them.

Today in the waters off Newfoundland, less than 1 percent of the cod that once swam those seas remain. And the majority of the surviving cod populations, elsewhere in the Atlantic, continue their steady decline. Cod remains the most widely eaten fish in France and most of Europe, a reliable staple, "guaranteed boneless." So, as if to continue the tradition of epic voyages, the last morsels of cod are flown to China in air-conditioned comfort, and then return packed full of additives and saddled with the weight of their story's sad end.

"Do you know what you'd like?" The waiter has returned. But we're looking for something heartier than cod, given the cool evening breeze. . . . Why not tagliatelle salmon? A creamy dish, delicious and filling.

An incessant plop-plop-plop noise hammers the water every day, echoing along the walls of the Norwegian fjord. It's like March hail, but all year round. A steady drizzle of granules drops into the confines of the aquafarm. The salmon doesn't have to worry about the daily menu. In aquaculture it's a constant rain of granules morning, noon, and night. But the salmon isn't especially hungry. Its instincts tell it to go in search of squid and anchovy, not to chase granules. Still, the granules are thrown at the salmon, again and again, granules seasoned with that entrancing odor of pheromones, driving the salmon to eat in spite of itself, driving it, along with the other 150,000 salmon in its enclosure, into a frenzied stampede to stuff itself full of insipid nutrients.

The granules sinking all around the fish now coat the bottom of the bay, and a fetid smell rises from the depths. The slightest cut on the salmon's fins, scraped on the nylon nets of its cage, will immediately become infected in the turbid water. It is sick every day of the week—except Tuesdays. On Tuesday the water tastes of antibiotics, and the salmon experiences a sudden rebound of good health. It never should have known instinctively what a Tuesday is. But for this salmon Tuesday is the fountain of youth.

When a westerly gale comes surging into the fjord, the aquaculture cage heaves with a sickening vertigo. One wave, higher than the others, triggers the salmon's reflexes and it shoots high into the air, falling outside the walls of its cage. Little by little, the salmon discovers a pure water it has never known, and as it follows the currents it meets strange salmon, fast and wild, and joins them in feasting on anchovy amid vast whirling eddies of shiny plankton and iridescent comb jellies.

But in all these new depths, there's no Tuesday. The salmon's old maladies come back to bite him, and soon the fluffy patches of fungal infection begin to contaminate the other salmon, who are defenseless in the face of these virulent fungi. Perhaps it will survive. Come springtime, it will feel the urge to swim up a river filled with memories, to find the stream of its birth and create new life there, in its turn. But this salmon wasn't born in a stream; it was born in a plastic basin. So it will wander, seeking the odor of its PVC birthplace, to no avail. It will wind up, discouraged, following the other salmon that *do* know where they're going. Perhaps it will even succeed in making its way with them past dams and nets and swim upstream, with the misgivings of an impostor, until it reaches some unknown river where, joining in the frenzy of communal love, it will spawn a new generation of equally disoriented salmon, who will surely never reach the sea.

"Ah, the tagliatelle, so sorry, we just ran out." The waiter sighs with feigned regret. Well then, the choice is made for you. "In that case, I'll have the roast chicken."

Off the coast of Peru, a net gathers up an entire shoal of anchovy over several miles, and with them a few dolphins and manta rays who were busy eating their fill. A sweep of 1,600 metric tons, that ought to bring a nice profit. No matter if the net crushes them, no one's going to eat them. Rich in bones and bitter-tasting fat, Peruvian anchovies aren't suited for our palates, but they have the advantage of living in large shoals, easy to target on the industrial scale. Back on shore they'll be ground into flour. To feed battery chickens.

DRAW ME A FISH

In which we try to get a fish stick to talk.

In which we develop a taste for anchovy.

In which the recipe for fish soup contains all the rules our ancestors imposed to protect the sea.

"Draw me a fish." At this instruction, given to elementary school children, most students took out a sheet of paper and sketched a rectangle. To them fish were golden rectangles, living in freezers, yet exhibiting a surprising range of biodiversity, because there's another species of fish too, one that takes the form of narrow sticks and lives in school cafeterias. A survey of 910 French children between the ages of eight and twelve shows that 20 percent have no idea that there's the slightest connection between the animals they see on TV, which are called "fish," and the food that appears regularly on their plates, which is called "fish sticks."

The stories of the sea fall silent in supermarket aisles, drowned out by street noise and silenced by thick cardboard walls of packaging. In cities the bond between human beings and the food they devour is broken. There is, however, a strong natural bond that typically links predator and prey, the bond of the food chain. In modern society we have lost our place in the food chain. We no longer catch what we eat ourselves. We only get to see our food under a processed form, turned into foodstuff, and over time we come to forget the very existence of the living beings we feed on. We deny the life of the creature that has become an industrial meal. We are deaf to the tales of the beings we consume, to all their stories, which are nonetheless every bit as fulfilling as the calories on the package. A fish on a sushi platter is no longer really "a fish." It is hardly even "some fish." It's nothing more than an abstract sliver. Between twin triangles of bread, the slim slice of salmon is bound and gagged. It can no longer tell us anything. We are busy. We gulp down our food without bothering to listen, rinsing it away with a swig of water after each mouthful to speed things up.

Anchovies on pizza can sometimes stir up debate, just like capers and olives. Some people love their salty flavor, others find them horrifying and pick them off. But whichever category you belong to, have you really given any thought to these anchovies? To their glittering schools of thin blue lines that quiver and ripple through the endless open sea? Everything depends on these anchovy schools. Dolphins, tunas, whales all feed on the immense biomass of these forage fish. A myriad of species owe their existence to the calories provided by anchovy.

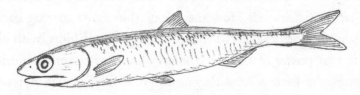

AN ANCHOVY

The anchovy has a strange head, with an immense mouth split all the way to behind the eyes, which helps it filter plankton and makes it look like a Muppet. It also has a large nose, which contains a sensory organ still not fully understood, the rostral organ, a sort of gelatinous mass filled with neurons linked to the optic nerve, thought to allow anchovy to pick up electric fields. Have you ever stopped to consider the lengthy history that connects us to the anchovy? For instance, the crazed appetite of the ancient Romans for *garum*, a sauce made of fermented anchovy, with a very "special" (a polite way to say "disgusting") taste, not unlike that of pure *nước mắm*, a strong Vietnamese fish sauce. For this awful *garum* the Romans were willing to fork over in sesterces the modern-day equivalent of 125 dollars a quart. Some historians claim it was a desire to control the critical strongholds of *garum* production in southern Gaul that persuaded Julius Caesar to invade France in the first place (except for a small village in Armorica, familiar to readers of the Asterix comic books). Have you listened to the anchovy's latest stories? How in 2005 the anchovy population of the Bay of Biscay nearly collapsed under the pressure of French fishing, only to be saved at the last possible moment by the mobilization and commitment of Spanish fishermen? How the population came roaring back once steps were taken to ensure its protection? How, with all the vast power of the sea, anchovy are now restoring life to the Bay of Biscay, attracting armies of cetaceans and myriads of birds? And how the same story is being repeated elsewhere across the world's oceans and seas, where more than 6.2 million metric tons of anchovy—amounting to every other anchovy alive—gets caught each year.

Society silences the stories of the fish, just as it silences human beings. We become shy due to indifference. We find ourselves in a world of overwhelming complexity, where everyone is scrambling to keep up, unable to stop. The boxes of hake in the frozen food aisle lie there like the hundreds of people in suits and ties in an office building. They've been dressed up, wrapped in a packaging of artificial color. Like all those workers, the hake has a part to play, one it never chose. And no one's going to ask it where it's from, or who it is. No one will let it speak, nobody will listen.

And yet if we take the time to listen to the ocean's stories, maybe we can help write them, choose their ending, play our part. I've heard a number of these stories in the depths of the sea. Some of them are quite sad. If you listen to the sea bass, gathered up by an immense open-sea trawl net along with its entire school in the middle of winter when it was about to spawn, captured in one fell swoop along with dozens of dolphins in a crushed heap on the boat's deck; if you listen to the 31 percent of the world's fish populations that are overfished and teetering on the brink of collapse, you'll hear their unhappy stories, stories in which faraway bureaucrats and unscrupulous lobbyists appropriate the sea and wring every last penny out of it. But there are also many lovely, cheerful stories. There's the line-caught saithe that die-hard handline fishermen from small Breton ports catch off the wild rocks, among the breaking waves, following the flights of seagulls. These fisherman are full of respect for the natural element that guarantees their livelihood. There's the Alaskan pollock, an industrially produced frozen square, whose cardboard box nonetheless bears a label stating that in the frigid seas where it was caught, scientists and fishermen have worked together to develop a fishery that is mindful of this natural resource. While mankind fuels an insatiable machine whose deranged course it can no longer control, men and women are finding ways to preserve life and restore the sea. They do so by inventing forward-thinking ideas for the future, or by revamping wise principles from the past.

In fish soup, or bouillabaisse, there is a blend of stories, if you're willing to listen closely. Every port on the Mediterranean will tell you that it alone uses the authentic recipe: extra saffron, less white wine, more aniseed, longer cooking time, leeks . . . I won't reveal my own recipe here for fear of starting a long and tiresome debate, and because a secret recipe should remain secret. But a proper fish soup—all the recipes agree on this point—must absolutely contain a great variety of rockfish. You'll need peacock wrasse, fine and rich in iodine; delicately flavored blacktail wrasse; ballan wrasse with its taste of seagrass, and scorpion fish with its powerful scents. At least seven species must be included, as Auguste Escoffier ("the king of chefs and the chef of kings") used to say, with all the authority of the inventor of crêpes Suzette and peach Melba. But plenty of cooks put far more than seven kinds of fish in their soup, happy to make good use of an unsold catch. This diversity reflects one of the principles of the small-scale coastal fisheries of the Mediterranean Sea, which have managed for the last five centuries to maintain a communitarian and ecologically sound management of this resource, in a virtuous and almost utopian way.

Since the fifteenth century, Mediterranean fishing ports

have maintained *prud'homies,* organizations of fishermen-judges elected by their peers and entrusted with the task of regulating coastal fisheries. Even before humanity realized that the Earth spins on its axis, they had understood that the sea belongs to everyone, and that its assets should be distributed fairly, on the basis of solidarity, fishing fleets, and equipment limits, with attention to seasonal catch diversification. The foundational principles were simple. Everyone must be able to eat his or her fill, and never take more than the sea can afford to give. Instead of destructive competition, local experts, working from common sense, set rules for sharing the fruits of the sea. All overly destructive equipment was banned. Fishing techniques and target species were varied, so as not to concentrate efforts on a single species, lest the ecosystem be disturbed and one fishery be given an advantage over the others. Fish soup thus contained a little bit of every species. This was a way not only to preserve fish populations, but also to give the soup an incredibly rich flavor. These structures still exist today, even if national and supranational interests have chosen to short-circuit them by underwriting industrial fisheries, which do not obey the same rules. The fisheries' *prud'homies* still foster their artisanal activity in accordance with principles from a bygone age. They have outlasted all the economic and bureaucratic pressures that attempted to subvert them. They keep fighting against all odds, thanks to their love for the sea and their traditions. You can still see the pointed and colorful bows of their boats in every Mediterranean port, living relics of an era from which we can all take inspiration. This memory of a harmonious past is also a source of hope, a forgotten seed that gave rise to principles which we may one day see blossom again.

I've been lucky to often witness the lives and secrets of underwater creatures. When their stories are told to us directly by the spectacle of the seabed, it's an immense delight to listen. But even when they come from a commercial product, a prepared dish far from the sea, it still can be a pleasure to hear them. Taking an interest in the origin of the product, in the living creature of whom we see a little part in the tray, and imagining where it came from and what sort of aquatic life it once led is already a step toward mending our broken link with nature. It's a first attempt to plot out our place in the food chain, to understand the role we play in it. And that intrinsically implies respect.

There is an entirely natural pleasure to discovering our place in the ecosystem again. Gathering sea urchins or shellfish revives our primordial instincts, which our brain then encourages us to follow. It resembles the simple joy of a child seeking Easter eggs, or a teenager's thrill in hunting for Pokémon. But by bringing it to bear in this way we can restore that joy to the original context in which our ancestors' brains first invented it, the food chain. Our natural instincts direct us to limit our take, to preserve the resources, to protect the secret of our fishing spot so we can return whenever we like and find fish there forever. Whereas supermarkets tempt us to buy more and more, nature encourages us to be moderate. By being aware of our place and role in the ecosystem, we preserve it.

It wasn't easy for me to find those roots, which struggle to penetrate the asphalt of big cities. Apart from the relatively remote stories we're told, can we glimpse the bond that unites us with other species when we're confined between streets and walls? I wanted to learn and tell the ocean's stories in order to share my love for it, but the alienating universe of the city kept me permanently at arm's length. In the city no one ever sees the earth except when construction work is under way; no one sees the sky except in patches between buildings. For the most part, people don't walk, instead letting themselves be swept along in the flood of mass transit, and we speak to each other from a distance, over electronic feeds and radio waves.

We even lose track of our place in space. The only orienting sense we have left is that of time. But often even time can't be perceived anymore, except perhaps in the form of stress.

When I was little, I often fantasized about the journey water made. I'd watch it flow away down the drain of the bathroom sink, and I could just picture those drops of water voyaging down the toboggan ride of the plumbing, racing toward the ocean. Water seemed to be escaping, down the sink.

If I dropped a line down the drain, I thought I might be able to unspool it all the way to the ocean, or at least some nearby river. And I pictured myself as an Inuit cutting a hole in the ice, hauling treasure chests or exotic fish, provided they were skinny enough to fit through the pipes. These miraculous fishing expeditions must have cost my parents at least a few rolls of twine.

I spent years in the city trying to regain my place in nature, attempting to grasp the stories of fish amid the noise of the streets. I had no idea just how close to me nature really was. I didn't have an inkling of the surprising discoveries I would make, just a few yards from my apartment, and the incredible species I would encounter beneath the concrete of the city's sidewalks and streets.

HOLD AN EEL
BY THE TALE

In which we open the door to an underwater Paris.

In which the most authentic Parisians are those who live under the Seine and have as many scales as clichés.

In which an eel wants to travel to the Caribbean so badly that it becomes immortal.

"Give me your credit card."
Fingers numbed by the cold, I slid the plastic card out of my wallet. A gloved hand gripped it in the half light. Dubious clicks and sounds of friction up and down the door. "Geez, it's really rusted shut." Suddenly the lock popped open. A screeching of hinges. "Let's go, guys, it's open now."

One after another, we melted into the darkness of the tunnel.

We were like three ghosts, surrounded by the mist of our breath, in the subterranean night. Once the door was closed behind us, it was time to switch on the lamps. A pale beam pierced the dark, outlining a circle of light in the water of the channel. The water was incredibly clear and calm. Without the silvery flames of the reflections dancing on the ceiling, you'd never even know there was any water.

The beams of our lamps wandered over the channel bed like paintbrushes, lighting a luminous window, unveiling its fleeting secrets. It was a stunning vista: valleys of bright sand strewn with shellfish and beer bottles, vast meadows of seagrass, heaps of drowned electric scooters. As we walked under the tunnel's vaulted ceilings, our eyes adjusted to the lamplight, following the rhythm of the beams as they swept the water. The light became a second line of sight.

All at once dozens of gleaming dots lit up in the shadows like reflectors on a highway at night.

"Look, there they are." One by one, the lamps went off.

Overhead the noise of traffic jams droned like a memory. The Métro rumbled occasionally, but it was no more than an echo that the sides of the tunnel turned into a melodious lament.

A few yards above us on the surface was the hustle and bustle of the streets, the asphalt coating the ground and the buildings reaching so high that they sliced the sky into slender slivers. There was the city, Paris. The city where I had so often felt uprooted, denatured. That city, which had long struck me as artificial, its sidewalks severing our vital bond with the earth, with life, with the elements. That city which, right now, I loved. Because I had finally discovered its submerged secret.

There are two kinds of Parisians: those who live underwater, and the rest.

I belonged to the rest, until one day I finally made the acquaintance of those who live underwater.

I met them with the help of a very strange brotherhood, the gang of Parisian street-fishers.

They are people like you and me, of all ages and backgrounds, but who, as soon as they have a few free hours, vanish into Paris's underbelly to explore a secret parallel world, armed with a headlamp and a fishing rod.

The fishing rod is only a pretext to get a closer look at the strange inhabitants of this universe, whom the street-fishers wrest briefly to light before returning them carefully to their element. Woe betide those who would damage the submerged ecosystems of Paris! The gang has connections everywhere, and it watches over the water's inhabitants as if they were family. Day and night, even at this very moment, they're omnipresent, under the streets, along the quays, in the forests and parks.

I was quick to join the secret expeditions of the street-fisher gang, and since meeting these underwater inhabitants I've never viewed Paris in quite the same way.

Like terrestrial Parisians, aquatic Parisians are Parisians first and foremost. They even share the same personalities, so typical of the French capital.

Aquatic Parisians are elegant snobs, especially in the nice neighborhoods. Along the quays near the Louvre and Notre-Dame are perches, classic Parisians. They match their chic striped raiment to the color of the Seine, and deck themselves out in red fins in the spring. Ever on the lookout for the latest fashion trends, they sneak glances at each other from the corners of their eyes, and if one seems to have made a good find, a cozy seagrass bed or a batch of gluten-free fry, the whole school will pivot to follow in an instant.

Under the sun of Paris-Plages, a yearly summer event in which artificial beaches pop up along the Seine, style-conscious fish work on their tans. These are the chubs. Elongated and silvery, they swim up the river against the current, so as not to appear mainstream. These freshwater hipsters change their diet daily. One evening they only have appetite for flying ants, the next day they've already turned vegan and will no longer deign to eat anything but dam moss algae.

Beneath the floating barge restaurants, you'll find the Paris of night owls. The wels catfish, for instance, never wakes up before nightfall, just in time to feast on scraps tossed from the portholes of the galley kitchens. This snake-like and slimy creature has such a ferocious appetite that it quickly grows to measure well over six feet. Like any authentic Parisian, it claims to come from a different region. It hails from the east, and its forebears swam in Lorraine back in the Ice Age. In those old days it must have developed a taste for Strasbourg sausages, which, used as bait, drive these catfish crazy. Practically blind, the wels catfish devours whatever prey it can detect with the help of its long, sensory whiskers. It is the scourge of Parisian waters, disturbing the sleep of both ducks and rodents.

But beneath its predatory, gluttonous veneer, this fish has strong family values. If you stroll along the banks of the Seine in June, keeping an eye on the roots of the weeping willows immersed in the water, you might just see a curious spectacle. Frightening-looking couples of wels catfish, black, scarred, and bewhiskered like sinister bikers, carefully take turns lingering in front of a delicate cradle made of algae and roots, and gently blow on their eggs in order to provide them oxygen, as the ass and the ox blew on baby Jesus. The male and female catfish guard their spawn for about ten days, until the young fry can swim by themselves.

When the river floods, Parisian fish also face the skyrocketing rents of the capital. They pack into the increasingly scarce areas sheltered from the current, under the Pont de la Concorde, or in the occasional suburban river bend. It's worse than trying to ride the commuter trains at rush hour. Teeming shoals of bream and common bleak cram together in the murky, muddy water, packed in with walleyes and pike.

Beneath the surface of the Seine and the Canal Saint-Martin there are crawfish who walk backward, and large freshwater mussels with mother-of-pearl shells in which bitterlings, a small minnow, lay their eggs. Goldfish abandoned by their owners regroup here, far from their fishbowls, and grow over two pounds in a few years. Thirty or so species of fish and hundreds of invertebrates inhabit this invisible and overlooked world. Every year new species recolonize these waters, as they become less and less polluted.

Some inhabitants are even more discreet. It's always night under the streets of Paris. In the dim light of the city's subterranean channels, the underground predators lurk in ambush.

Pairs of luminescent eyes glittered just out of reach of our lamps. We ventured cautiously closer.

The first shapes to appear in the half-light waved silently, as if they had emerged from a dream. False alarm. These weren't what we'd been looking for, but instead slow-moving eels swimming in the lamplight, serpentine with their shimmering skin. Anyone who has ever seen an eel quickly realizes that these fish are not insignificant, and that they conceal mysteries beneath their strange, hybrid appearance.

The eels of Paris, like all the eels of Europe, were born in the Caribbean. No one knows the exact location of their birthplace, but it's commonly thought to be near the Sargasso Sea, to the northeast of the West Indies, probably at incredible depths. Leptocephali, or newborn eel larvae, are only a few millimeters long and resemble a willow leaf. They are so transparent that with the naked eye you can only see the plankton that moves in the wake of their wavy wiggling. They are equipped with dragon teeth, gigantic in relation to their size. To reach the coasts of Europe and North America, several thousand miles away, they swim through the Gulf Stream for months on end, without stopping. Along the journey the leptocephali gradually transform, to take on their final serpentine appearance. When they reach the mouth of a river, they swim upstream in the form of glass eels, or elvers, miniature eels already. Passing from salt- to fresh water causes an osmotic shock that would prove fatal to most fish, but this is the least of the eel's exploits. Once an eel decides to swim up a river in search of a quiet stretch of riverbank to inhabit, nothing can stop it. If the river is dammed, the eel will climb and crawl over land and fields, for several days if necessary, holding its breath. If it can't find any open water, it will slip through any pipe, conduit, or resurgence it can find. It will even wriggle through subterranean water tables until it can reach a stream.

Once it has found one, the eel recovers its strength and grows, until the day it hears the sea beckon, after about ten years of quiet freshwater life. Answering the call of the sea, the yellow-skinned eel adorns itself in a silvery livery, swims downstream to the estuary, and sets out for the depths of the Sargasso Sea of its birth, where it will love and where it will die, perishing while producing new life, as part of a dark mystery. In spite of more than a century of research, no one has ever been able to follow the eels to the end of their journey, nor to discover the exact place where, after six months of swimming without stopping or eating, they give life to a new generation.

Why do eels stubbornly persist in making such a long journey to lay their eggs? It's an old story, older than the ocean itself. Millions of years ago eels spawned near the coasts of Europe and America. Back then the Atlantic was nothing more than a small young sea, and Europe and America were close neighbors. But little by little, continental drift drove Europe away from the American continent at a speed of several centimeters per year. Eels from Europe and America didn't notice what was happening, so they went on traveling to lay their eggs in the one place where temperature and seabed were ideal for their purposes. They remained faithful to the waters of their birth, and as these waters moved farther and farther away, they adapted to make that long journey, which now takes them thousands of miles. These are some extremely tenacious fish.

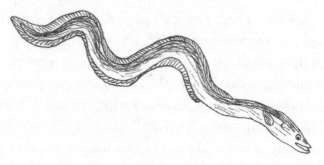

EUROPEAN EEL

If by some mishap an insurmountable obstacle prevents an eel from reaching the ocean when its fate calls, it is prepared to wait patiently, for all eternity if necessary. Trapped in fresh water with no possible access to the sea, an eel will abandon its silvery traveling clothes, put back on its golden livery, and wait as long as it takes for the obstacle to disappear, as if it were immortal. As long as it's kept from its fate, the eel seems unwilling to die.

In the Swedish village of Brantevik, in 1859, the year Victor Hugo was writing his *Contemplations,* Samuel Nilsson was just eight years old, and he tossed an eel into the well of his grandparents' home. Tossing an eel into a well was not considered an especially naughty thing for an eight-year-old to do. In fact, an eel could be a very good way of cleansing the well of insects and other vermin that might pollute its water. Samuel's grandparents didn't hold it against him, and even decided to let the eel live in the well. The misdeed seemed to be forgotten, but Samuel would never have imagined that his own great-great-grandchildren would still hear talk of it. He named the eel Åle—not an especially original name, considering that it means "eel" in Swedish.

Since the well had no outlet, the eel had no way to get back to the sea, and resigned itself to a long wait. The months turned into years. Samuel Nilsson grew up and left his grandparents' house, and by now Victor Hugo was writing *Les Misérables.* Little by little, Åle's eyes adapted to the darkness. The years turned into decades. The house changed hands, one generation succeeded another. Victor Hugo was buried in the Panthéon, mankind invented the automobile and then the airplane, there were two world wars and various nuclear catastrophes, and Neil Armstrong walked on the Moon. The eel, meanwhile, remained in the well, waiting. There were great discoveries and revolutions. Åle still haunted the well and was even featured more than once in the local news column of the Brantevik newspaper. One day a fellow eel was tossed into the well to entertain him. In Japan eating imported eel larvae from around the world became a luxury trend. The species, once so abundant that it was considered a pest, began to decline in number across Europe, and eventually eels were classified as a critically endangered species. Some 90 percent of the eel population vanished. But Åle was unaware of all this. He had made up his mind to go on living until he found a way out of that well to reach the Sargasso Sea. Time meant nothing to him.

This eel's story ended tragically during the traditional Swedish crayfish party in the summer of 2014. Because of a poorly insulated well cover, the water heated up, and Åle was found boiled to death. He had lived to the ripe old age of 155. His fellow (or sister) eel, now 110 and still unnamed, survived him, and still bides her time patiently in the well. That eel has waited for more than a century. Has she developed a taste for immortality, or has she merely survived in the hope she might someday fulfill her destiny? If this eel were now let loose into open water, how would she feel? Released into a brand-new world alongside the very last surviving members of her species, would she feel a surge of joy as she swims away from the eternity of the well, setting off on a one-way journey to the Sargasso Sea?

We hadn't come to observe eels that night in the canal.

So our lamps began moving again, sweeping across the gravel bed. Beneath the carved stone banks were hundreds of Asian clams, little shellfish that gave the slabs they clung to the texture of plaster. Ruffe fish, almost translucent creatures with spiny fins, darted through the water; we could see their retinas gleam like will-o'-the-wisps in the lamplight. Pale minnows slept twitching in midwater. Sometimes we happened upon a bronze carp that scuttled off in a placid flop of the fins. Our eyes could make out nothing but the circle illuminated by our lamps, in which the shadows of fish were outlined like the silhouettes of actors in a spotlight. We advanced like sleepwalkers amid the cold and the underground echoes. Bats squealed against the ceiling with the sound of pencil sharpeners. A heron leapt into the air from the opposite platform and vanished like a ghost.

"Down there, look, I see one!" Two pearly retinas, large and round, shone in the pitch black, and I thought I could just make out a squat brown shape. The luminous eyes slowly moved away.

It was a zander, a nocturnal predator, exactly the fish we were looking for. Midway between a pike and a perch, the zander is a kind of big dark walleye with a set of sharp teeth—an elusive, savage predator. We were slow and steady in our approach to keep from startling the zander, never letting the circle of light brush its shadow as it gradually withdrew into the watery darkness.

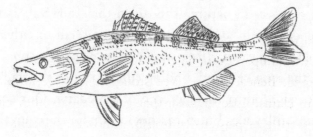

ZANDER

Pursuing that fleeing chimerical shape in that subterranean universe filled me with an emotion that was at once strange and too intense to be entirely unfamiliar. I felt that primitive, animal satisfaction, a sensation of being completely alert—my gaze, my pulse, and my thoughts all wholly absorbed in the observation of nature, at one with the water and with life. Looking out for any sign of the fish's presence, trying to anticipate its movements, I was no longer myself but a predator hunting its prey. I thought again of that piece of string I had uncoiled down the sink drain as a child, hoping it would reach open waters and return with a fish at the end of the line. That dream hadn't been so crazy after all; wildlife wasn't all that far away. It was still hiding, but it was waiting for me. In a tunnel thirty feet below the concrete of Paris, exploring an underground canal, I'd reconnected with that thread again, found that original place in the great chain of life.

Originally evolved to survive in nature, we're hardwired to be at our happiest while stalking our prey and when we escape from our predators. In such situations the corpus striatum of the human brain rewarded our ancestors with dopamine, a genuine happiness drug, for any useful clue they might discover in nature. This aided their survival, as they sought out joy in anything that allowed them to either eat or avoid being eaten. These were our ancestral pleasures: noticing the song of a bird, finding edible fruit or the tracks of prey, or else foiling the approach of a predator. Our corpus striatum still works, but it searches in vain for the equivalents of those primitive joys in a modern world that so thoroughly disorients it.

"Look out, here comes the River Brigade!"

Did I get a jolt of dopamine when two spotlights, the barking of a Belgian shepherd, and the sound of booted footsteps erupted from the mouth of the tunnel? Whatever the case, in an instant, I took my rightful place in the food chain. Predators for the unsuspecting fish, we had in our turn fallen prey to the Paris gendarmes.

Absorbed in our aquatic Paris, we had forgotten the "No Trespassing" sign hung by land-dwelling Parisians at the entrance. The door whose lock we'd picked was already far away. But the instinct of the hunted animal is quick to react in such a situation. A surge of adrenaline mixed in with the dopamine, and we were running for the exit.

The debacle was anything but glorious, but our pursuers couldn't keep up with our primitive animal instinct. We escaped without their ever realizing we were there, in a place that was strictly off-limits, for no other reason than to look at fish.

Back out in the fresh air, walking in the yellow glow of the streetlamps, our misadventure was already taking on the glamour of urban legend. "That dog didn't look happy, and it wasn't wearing a muzzle either! Lucky we didn't trip over the cables." "How much do you think the fine would have been if they'd caught us?" "I have no idea, and I'd rather not find out!" "It really is exhausting to run in the dark." "That zander must have been twenty pounds, I've never seen one like it." "What about the eels, did you see those eels?! Some of them were as thick as my leg." "If it weren't for the cops, we could have caught the zander. It would have made for a great picture."

Already our adventure was taking on the glow of romance. The tunnel was growing darker, the zander bigger, and the gendarmes scarier. The fish hiding in the canal were starting to tell their tale through our panted exclamations.

We hadn't caught the zander, but we'd caught a story.

SEA SERPENTS

In which nonexistent species also deserve our protection.

In which we remember the remora of the Romans.

In which the sea serpent exists, and can predict earthquakes.

What did the underwater world look like ten thousand years ago, before the first diver ever saw it? We can easily imagine the earth before civilization, covered with dense forests and wild steppes, without cities or roads, without power lines or cultivated fields.

But the sea, what did the sea look like?

No doubt the sea was far more populous than it is today. Monk seals, members of the Phocidae family, abounded on every Mediterranean beach, and were once so numerous in the Turkish islands that they gave their name to the Phocaeans, whose voyagers settled on the shores of what is now Provence, in the city of Marseilles, still known today as the "Phocaean city."

There are no more than five hundred monk seals left today, hidden in the seclusion of remote caves.

In the Bering Strait vast herds of giant manatees grazed in seaweed prairies as recently as three hundred years ago. These animals, known as Steller's sea cows, grew to lengths of up to thirty feet. The last one was hunted down two hundred years ago. As I write these words, there are no more than ten surviving vaquitas, the smallest living cetacean on earth, a diminutive black-and-white porpoise found in the Gulf of California. On a global scale the populations of large fish have fallen by two-thirds over the course of the last hundred years.

This fact is alarming, to say the least. We have every reason to be concerned about the extinction of species that exist today. But who worries about the extinction of creatures that don't exist? They, too, are critically endangered!

What has become of the sea serpents who, according to the sailors of bygone eras, were once so plentiful that they caused shipwrecks? Who among us in the last two centuries has heard the siren's song? Where are the mermen, the krakens? Though these animals may never have existed, are they, too, on the verge of extinction?

I wish I could have seen the prehistoric seas, because they were inhabited not only by vast numbers of fish but also by stories.

Despite its name, the prehistoric era was when stories were first born. Before there was a way of writing, or of keeping records, these stories could only live in the imaginations of those who heard them, and so they were transformed constantly as they were handed down, as free and ephemeral as words themselves.

In order to explain and envision the world, people believed all the legends; every stretch of water was filled with myths and imagination. The sea sailed by prehistoric men abounded with improbable creatures, supernatural entities, phantasmagorical beasts. It was a primordial ocean populated with fantastical beings.

But one day around 3400 BC, humans invented writing. Scholars appeared, and they decided to record what they knew about the creatures of the sea, in hopes of better understanding them. We have them to thank for their invaluable accounts of these bygone beings, but we also have them to blame for the disappearance of many of those creatures, which they deemed merely fanciful. Such creatures, these scholars simply decreed, no longer existed.

Pliny the Elder, an ancient Roman philosopher and a high official in Transalpine Gaul, later known as Occitania, aspired to collect all the knowledge of his time in his *Natural History,* written in AD 77. Volume IX is dedicated to the sea and offers a delightful overview of the creatures that populated the waters in ancient Roman times. According to Pliny, there were precisely seventy-four species of fish and thirty species of crustaceans, and he seemed quite confident in those numbers!

To write his magnum opus, Pliny claimed, he read more

than two thousand books, written by over five hundred authors, to which he added his own personal observations. Apparently his work schedule as a high functionary in Gaul left him plenty of free time. He accepted the accounts that struck him as serious and let the others fall into oblivion. Certain sections of his *Natural History* have been confirmed by modern-day science. For example, he accurately reported, roughly two thousand years ago, that the comber, a Mediterranean rockfish, is a synchronous hermaphrodite fish species, meaning it is both male and female. He also understood that the electric ray is ovoviviparous and spawns eggs that hatch in its womb. He further noted that seals enjoy a very deep sleep. Nowadays we know that these mammals are, like human beings, capable of experiencing REM sleep, as well as dreaming. Pliny, who had no understanding of modern neurology, believed it was the seal's left fin, placed beneath its head during nap time, that possessed soporific virtues.

Pliny also wrote other, more fanciful passages, reflective of the beliefs and the science of his time. He added his own dash of spice to his descriptions of the behavior of the remora, a fish with a sucker-like organ on its head that latches onto larger fish to exploit them for free rides and scraps from their meals. According to Pliny, this clingy fish possessed the power to slow down the ships to which it attached itself, even halting them entirely. For all the mariners of his time, this was obvious, and in Latin, "remora" was the word used for "delay." On the second day of September in the year 31 BC, the decisive battle of Actium took place, the outcome of which would determine whether Marc Antony or Octavian would succeed Julius Caesar to become the emperor of Rome. Marc Antony's fleet was superior in number, and theoretically the larger fleet was bound to win. But under the effect of some mysterious force, Marc Antony's galleys were stopped in their attack and slowed to a halt. This incident gave Octavian a strategic advantage, ultimately leading to his victory.

According to Pliny, it was obviously the remora that had caused this unexpected turn of events.

Some of the fish Pliny described were quite similar to those we know today, such as the bluefin tuna, whose maximum weight is estimated to reach one thousand four hundred pounds. (The current official game fishing record for the fish, caught off Nova Scotia in 1979, is 1,496 pounds).

But in Pliny's sea there were also whales ten leagues in length, so big that they couldn't move without creating storms, as well as sea turtles from the Indian Ocean whose shells were large enough to serve as the roof of a house. Pliny also claimed to know the true origin of the mermen who played conch shells at deafening volume in undersea caves. No one in living memory has seen them.

As scholars wrote page after page of bestiaries, new creatures were discovered and old ones forgotten. But the more knowledge advanced, the more those legendary beasts withdrew into silence.

While the bestiaries of the Middle Ages were still filled with sea monsters, sadly their whales no longer stretched as far as the eye could see. On the other hand, there were sailors who allegedly mistook them for islands. They would moor their ships and set foot on those whales, thinking they had reached dry land. But they always seemed to wind up lighting a fire on the whale's back, triggering its rage, and the whale would dive, dragging with it the entire cargo and crew, body and soul, into the deep.

At the time there was no easy way to share images. No one knew how, nor was it possible to make a sketch in the moment. For that you had to have a quill and an inkpot, and also a table. Sketches and descriptions were all deformed by memory and narrative, which meant the seas were still populated by beings enhanced by the imagination.

Still, knowledge was gaining ground, and with it came a certain rigor. Naïvely believing was no longer enough; now people were starting to demand proof. We got rid of mermen, for lack of proof of their existence. Whales remained cloaked in superstition, but soon no one seriously believed a cetacean had ever been mistaken for an island.

The sciences grew more formalized, and in the eighteenth century taxonomy developed. Before a species could be officially recognized, it had to be given a scientific name in Latin and Greek, and specimens had to be collected as proof. The

Swedish botanist Carl Linnaeus was one of the first to catalogue thousands of species in accordance with official protocol. His vocation as a pioneer of the field of nomenclature was ironically predetermined, since Linnaeus himself changed names no fewer than nine times over the course of his life, finally borrowing his name from another species, the linden tree that grew on his family farm. ("Linnaeus" in Latinized Swedish.) By 1758 Linnaeus had classified in his *Systema Naturae* some 4,400 animal species and 7,700 plant species, naming them according to the binomial taxonomic system that is now used for all living things. He catalogued each species on an index card and gave it a place in the tree of life. If there was too little evidence of its existence, he'd eliminate it from the map. Sea monsters therefore disappeared en masse. There was a full-fledged massacre of imaginary beings. Any creature unable to provide supporting evidence of its existence no longer even had the right to a name. They were denied the right to existence. Did Linnaeus himself feel any remorse about the reckless pragmatism of his undertaking? The scientific nobility of his project no doubt blinded him to all else. But when the time came to give a name to the immense blue whale, the largest animal of all time, this creature that had unleashed so many myths and exaggerations, the austere Swedish savant couldn't refrain from giving the last word to his imagination, indulging in a bit of fun. He called it *Balaenoptera musculus,* which in Latin means "mouse whale."

As for the remora, Linnaeus decided to exonerate it of any involvement in the Battle of Actium. He had no idea what had actually stopped Marc Antony's fleet, but he felt quite certain that a fish measuring no more than a foot long couldn't slow down a galley. He stripped the remora of its magical powers, but magnanimously he left the remora a hint of that power in its name, *Echeneis naucrates,* which means "that which holds back ships."

It would take until 2018 for a team of physicists, through a series of calculations and advanced simulations, to solve the mystery of the remora and discover the actual cause of Marc Antony's defeat at Actium. A sudden change in the water's depth along the shoreline caused a rare hydrodynamic phenomenon, a solitary wave that prevented the fleet from advancing, halting it entirely.

The great oceanographic campaigns of the nineteenth century managed to eliminate the sea monsters of olden days, sea monsters that had never existed but nonetheless populated the aquatic imaginations of sailors over the centuries. Documentary sketches became precise and realistic, and were then replaced by photographs. Today the research vessels that explore the ocean floors record the DNA sequences of living beings without even seeing them. They capture plankton samples in special trawl nets and then undertake a "massive parallel sequencing," which gives them access to the entire genetic history of the creatures they've captured. Scientists have observed and described fish all the way at the bottom of the Mariana Trench, at a depth of 10,900 meters, the lowest point in the ocean. The giant squid that once sank ships in sailors' dreams and drawings have now been filmed and measured. We can observe whales on any screen at the touch of a button from the comfort of our sofas, and confirm that they look nothing like floating islands.

Where can we still dream, now that the monsters have been chased from the world's imaginary, and the depths of the sea are revealed to us in high-definition, full-color videos? How can we go on creating stories?

After all, we have a deep and abiding need to believe and to dream.

One day on a beach in New Zealand, I was taking a swim like a good tourist, between offshore trips in search of yellowtail amberjacks, when I happened to spot two bluish fins breaking the water's surface in the middle of the swimming area. I ventured closer, thinking it was a ray. That's when I realized it was actually a blue shark, a species I knew quite well because it's also found in the Mediterranean.

These peaceful animals, magnificent in color, generally live in the open water. This one, clearly bewildered by the shallows into which it had wandered, wound up beaching itself on the sandy shore. To keep a memory of the encounter, I filmed the scene as I guided the shark back out into the open water, where it swam off into the blue. I posted the video on the internet to share with my friends.

It was quite a surprise, several months later, when I found an article published by *Daily Mail Australia*, with the following caption beneath a screen grab from my video: "A fearless New Zealand man grabs a man-eating shark's fin with his bare hands." In the wake of the excitement stirred by the video, a journalist—blithely ignoring my caption, which was detailed, but in French unfortunately—had proceeded to invent from whole cloth a thrilling tale in which this mild-mannered shark was portrayed as a bloodthirsty beast. The article was of course met with a storm of online commentary, spurring fascinating debates. One commentator exclaimed, "He's a criminal: he let the shark go, and it could have come back at any moment to eat innocent children!" And another retorted, "Sharks are endangered species, and children are not, so to me he's a hero."

The blue shark is a species that feeds on small prey such as anchovy, and never attacks humans. So I contacted the journalist, revealing the truth and asking her to set the record straight. She corrected only the expression "man-eating shark," changing it to "potentially man-eating shark." To which I responded that she was "potentially" a good journalist. (She may have potentially grasped the irony.) But she stood firm on her fear of sharks.

Why do people today always seem to be afraid of sharks, when statistics tell us that toasters kill more than ten times as many people each year than all sharks combined? Surely it's an age-old need to confront something more powerful than we are, to be able to feel insignificant in the face of nature and its supernatural might. We no longer have any predators, so we're excited to contemplate the possibility of one. This imaginary predator, this creature larger than us that just might hunt us, gives us a taste of the place we've lost in the food chain and nature's cycles. For lack of sea monsters, we invent them.

We've banished monsters from our modern bestiary so thoroughly that nature seems to want to prove that we'd be better off believing in them. In fact, reality sometimes outstrips legend.

Take the sea serpent, which has haunted seafaring mythology for centuries. Scientists eventually came to the conclusion that they don't exist, that they were merely cooked up by the overactive imaginations of sailors, and that the only snakes adapted to life in the sea never grow over seven feet long. Then one day the sea finally decided to present scientists with a sea serpent, just to prove they really did exist.

GIANT OARFISH AND SEA SERPENT

The giant oarfish is an incredible creature—a fish shaped like a serpent and reaching up to thirty-five feet in length. The fish is silvery with blue spots, topped by a long red dragon's crest. Observed only rarely, it perfectly matches most recorded descriptions of sea serpents, and no doubt inspired people to imagine them in the first place. But what we've learned about its life actually far surpasses the most implausible legends. We recently discovered that the animal can swim backward, and vertically as well. It also practices autotomy, or self-amputation, meaning that it can cut itself in two and leave behind a part of its tail to escape a predator, or just to save energy by reducing its size. We might even suppose that, in case of famine, it could eat itself, like the notorious snake from the video game. Even better, the giant oarfish is thought to be capable of predicting earthquakes. There's no other explanation for the troubling correlation, observed throughout the world, between the beaching of a giant oarfish and the occurrence of an earthquake. Both events are quite rare, but they frequently coincide. The giant oarfish seems to live near oceanic faults, and would appear to be mysteriously sensitive to their activity.

I have yet to see a giant oarfish in the flesh, but one day a friend informed me that one had just washed up near Cannes and showed me the video. We happened to be on a boat at the time, and we joked about the earthquake we would escape. We were stunned that same evening when the local news informed us that a small earthquake had in fact been detected in the region. The epicenter was determined to have been a few miles off the rocky shore where the giant oarfish had washed up.

According to the most recent estimates, there are 2.2 million species living in the sea, not including the billions of species of bacteria. Humans have identified less than 10 percent of them. We model the number of species yet to be discovered by analyzing the already well-known groups of living creatures, and the rate at which they have been discovered in years past. While we no longer believe in the legends described by Pliny, we can still safely state that our current knowledge is infinitesimal, and will turn out in the future to be as inaccurate as past knowledge has turned out to be in our time.

One day in the future, people will laugh at the certainties of our era, just as we now chuckle at the beliefs of the past, such as the staunch conviction that the Earth was flat and that the seas were inhabited by no more than seventy-four species of fish.

In the meantime, roughly 91 percent of all marine species remain unknown, leaving us myths still to be written, blank pages with room to dream. In the darkness of the seas swim the future discoveries that simply await our imagination to exist and inspire us with their legends.

We are free to believe in these dreams, to listen to these stories, to give them life. And—why not?—to bring back certain ancient legends. The sea serpent did exist after all.

THE SEA IS YOUR MIRROR

In which our world is reflected in the sea, which
is a reflection of our world, which is a reflection of
the sea.

In which migratory geese are born in crustaceans.

In which a jellyfish wins two Nobel Prizes.

In which the sea is reflected in our world, which
is a reflection of the sea, which is a reflection of
our world . . .

In the imaginations of yore, the sea teemed with mythical
creatures primarily due to a very ancient and persistent legend: the legend of the mirror.

Have you ever wondered why so many sea creatures bear
the names of terrestrial beings?

You can find all the animals of Noah's Ark under the sea: catfish, lizard fish, scorpion fish, dogfish, squirrelfish, wolf fish, parrotfish, rabbitfish, toadfish, sea hares, sea elephants, sea lions, sea cows, sea pigs, sea leopards. There are even sea grapes, sea tomatoes, sea apples, and sea cucumbers. The most astonishing array of objects have a marine counterpart: sea stars, rockfish, sawfish, moonfish, sunfish, trumpet fish, boxfish, balloonfish. The same goes for professions: monk seals, cardinal fish, clown fish, soldierfish, surgeonfish. And even divine beings such as sea angels, sea devils, unicorn fish.

These names all originate from the ancient legend of the mirror. Our ancestors believed that beneath the mirror of its surface, the ocean was a parallel world, conceived as a mirror of the earth. Everything that existed on dry land necessarily had its equivalent somewhere under the sea.

This ancestral theory likely came about naturally in prehistoric times. It's such an obvious thing, when looking at the sea, to glimpse your own reflection. To see the colors of the sky displayed in it upside down, to see fish swimming in it like birds flying through the sky.

Pliny picked up on this popular belief. He also noticed, on the beaches of the Mediterranean and in the tales of travelers, the raisin-like eggs of cuttlefish, the saws of sawfish, the swords of swordfish, and the cucumbers of the sea. He was astonished by the similarities, as if sea creatures were slightly modified copies of beings on land, as is the case with the seahorse, with its horse's head at the end of "a small snail." To explain these observations, he proffered a hypothesis. The seeds and embryos of beings, rolled around by the waves and the winds, mingled between the air and the tide, resulting in this strange hybridization between the inhabitants of the two worlds.

Over time books were copied and legends perpetuated. This belief spread and permeated the minds of all Europe.

In the Middle Ages monastery scribes took Pliny's ancient observations quite literally. The legend of the mirror became a cosmological concept. The most famous medieval scholars, whose names alone are a chivalric saga—Godfrey of Viterbo, Thomas of Cantimpré, and Gervase of Tilbury, for instance—wrote that the sea was a world parallel to our own, and that each terrestrial being necessarily had its counterpart beneath the waves. According to Gervase, a land creature's marine equivalent resembled it "from the head to the navel," but its body often ended in a fishtail. And this undersea world must have had its animals and plants, but also no doubt its civilized peoples, to match the humans on land.

As soon as a new sea creature was observed, the greatest minds set to work to identify the land animal whose aquatic counterpart had just been glimpsed.

Thus the swordfish and its blade were thought to be the sword of a marine knight, whose shield was the sea turtle, and whose helmet was made of the largest crabs.

The illustrators of the time had no idea how to draw fish, though they had plenty of models of land animals to work from. So they painted creatures that closely resembled the animals found on dry land, then simply tacked on fishtails, and the illustrations they provided for bestiaries perpetuated the legend of this imaginary subaquatic world.

The church supported the mirror theory because it high-
lighted and reinforced God's creative power. It's worth men-
tioning that this theory also allowed for certain fantasies and
configurations that were rather creative, to say the least.

Along the rocky coasts of the Atlantic and on planks of driftwood there are often barnacles—sedentary, clinging crustaceans that look like a sort of pale-colored mussel in the shape
of a bird's beak at the end of a short black tube. This species
was abundant in the twelfth century along the European
coasts, but no one was able to identify a land-based equivalent.

The British clergy leapt at the chance to provide a response
as cunning as it was unexpected. It was the end of winter,
the time of famine, and priests and burghers were starting to
crave the taste of meat, forbidden throughout the forty days of
Lent. In that season barnacle-covered driftwood would wash
up on the beaches of northern Europe. At the same time, in
the sky overhead, small black-and-white geese called barnacle
geese were beginning to fly north in their annual migration.
These birds vanished to nest in some unknown location. Back
then no one had ever traveled to the Svalbard archipelago,
north of the polar circle, where they reproduce.

The Welsh monk Gerald of Wales had an idea. He saw the
barnacle geese setting off for who knows where, he saw the

barnacles washing up out of nowhere, and he was suffering from the privations of Lent, which kept him from savoring a nice fatty meal, so he decided to solve all three problems in one fell swoop. He wrote that their tube-shaped "neck" and shell "beak" proved that barnacles were the young and still immature "sprouts" of the barnacle geese, and proclaimed his discovery far and wide. All the experts of the time conceded in concert that these crustaceans were surely the marine equivalent of the barnacle goose, and even better, that they would transform into geese as they matured in the high seas. The specimens washed up on the beach were not yet mature, having only a fully formed beak, but they would later grow feathers, then wings, and finally take off as full-grown barnacle geese. Therefore the clergy of the time decreed that the goose had been born of the barnacle, making it seafood, which meant that the Europeans of the Middle Ages were entitled to dine on goose flesh even during Lent. The crustacean was christened *anatifer,* which literally means "goose bringer" in Latin. This later became *anatife* in French. In English the same word, "barnacle," is still used today to describe both the barnacle goose bird and the goose barnacle crustacean.

This belief was slow to die. Rabelais mentions it in *Gargantua and Pantagruel,* and until the turn of the nineteenth century barnacle geese were considered seafood in Scotland. The monk Gerald had no idea that the true story of the barnacle goose, though it had nothing to do with the crustacean in question, was no less extraordinary. These birds nest in the shelter of the high cliffs of the Great North, a height they reach after a dangerous four-thousand-mile migration. They

GOOSE BARNACLES AND BARNACLE GEESE

never get lost. Almost immediately after a gosling hatches it must throw itself into the void from high atop the cliff, to reach the water and the moss of the tundra, without even knowing how to fly. This often means falling as much as four hundred feet, with this tiny ball of goose down bouncing violently off the rocks. However, generally speaking, the gosling is sufficiently soft and light to survive the terrible ordeal.

In our ancestors' imaginations, the sea long remained a mirror of the terrestrial world, even well past the Middle Ages. The appearance of unknown species often aroused suspicions. Were these signs of the existence of an underwater civilization? In 1551 the naturalist and physician Guillaume Rondelet described having seen a "sea bishop," a sea monster wearing a bishop's cassock, in the North Sea. Presented as a curiosity to the court of the king of Poland, the sea bishop displayed such eagerness to return to the waves that its wish was soon fulfilled, and as it swam away it bade farewell to its human captors by making the sign of the cross. The Renaissance men believed they had just seen an ambassador from a subaquatic civilization. In all likelihood, however, it was a hooded seal, a species whose males have a red stretchy cavity or hood on their head, and who can move their fins in a way that resembles hand signals. Unless, that is, the sea bishop concealed some other strange and suspicious animal, whose dried and cured skin was sold by countless opportunistic charlatans to cabinets of curiosities with the dubious claim that it was really the dress of an underwater clergyman. The creature was a large cartilaginous fish. It was mistaken for a bishop in northern Europe, while southern Europeans took it for an angel. Indeed, in Nice, up until the twentieth century, immense fish with flattened wings and rough skin, somewhere between a ray and a shark, frequently destroyed the nets of sardine fishermen. At the sight of their large wings, the fishermen dubbed them *lu pei ange,* "the angel fish" or, as we now know them, "angel sharks." These *Squatina squatina* are now quite rare along our coastlines, but there is a place between Nice and Antibes that still commemorates their presence, the Bay of Angels.

Today no one believes in the legend of the mirror anymore. We know now that life originated in the oceans, in a community of bacteria that gradually diversified into an array of plants and animals. Certain creatures evolved and left the ocean's depths for the open air, where oxygen is more concentrated than in the water. These creatures colonized dry land. Other species remained in the water, taking various forms and shapes. Others still, such as the ancestors of the cetaceans, left the water, adapted to life on land, and then returned to the sea, once again taking on fishlike traits in order to adapt to their environment through a mechanism known as convergent evolution.

The legend of the mirror has been forgotten. It has joined the bestiary of ancient and superannuated superstitions, filed away alongside the engravings of whales so huge they were mistaken for islands and melodies once played on flutes to keep sea monsters at bay.

But now science and technology have taken over in giving life to that legend, or rather, to the reflection of that legend in its own mirror. Nowadays—and this is a forward-looking trend—our terrestrial world takes inspiration from the underwater world and does its best to emulate it, as in a mirror.

Evolution forged the hammerhead shark millions of years before the first hammer was made, and the first mother-of-pearl shell long before the first composite materials were concocted on land. In the 3.5-billion-year history of life, nature has developed and tested, by means of natural selection, countless technical solutions to ensure the survival of its species, creating a trove of invaluable sources of inspiration for inventors in every field.

Little by little, our terranean world is becoming a mirror of the seas, imitating them in the form of so-called biomimicry, or biomimetic inventions. For example, the corrugated metal of our roofs is an imitation of the particularly strong and resistant structure of the scallop's shell. The fuselages of many of our vehicles are modeled on the hydrodynamic shapes of fish. Surgical robots imitate the supple agility of an octopus's tentacles. Countless medications are copies of marine molecules, from the poisons of sea cone shells to the proteins of sea squirts. Nature often outdoes engineers, and it offers them a cornucopia of models to imitate.

Deep-sea Euplectella sponges, also known as glass sponges, can construct a glass skeleton without any need for a high-temperature kiln or industrial chemistry. Their glass possesses exceptional optical properties as well, far better than our own fiber optics, and they use it to intensify the light of bioluminescent plankton in order to attract and feed on planktonic algae. These strange sponges can live for up to thirteen thousand years. Young shrimp couples set up housekeeping when they're still very young in the skeletons of the Euplectella sponges, which are braided like baskets. As they grow larger, the shrimps discover that they can't get out, and so live the rest of their lives together, housed by the sponge. Aside from the sponge's value as a symbol of fidelity, a number of future inventions are likely to stem from studies of Euplectella's material. Research is already being done on them to model architectural structures, as well as biocompatible prosthetics and innovative types of glass.

Technical revolutions are already emerging from the oceans. The hemoglobin of lugworms, sea worms that litter beaches with their sand coils, transports oxygen forty times better than human hemoglobin, and works for all blood types. This model has inspired recent products that preserve organs for transplants roughly ten times longer than the customary solutions.

Some of these aquatic inspirations have had stunning consequences for our world.

The *Aequorea victoria* jellyfish lives in North American waters, where it feeds on copepods and other species of jellyfish. In order to attract its prey, it produces an intense green light by means of a fluorescent protein. When it succeeds in capturing a jellyfish roughly half its size and, with ravenous satisfaction, expands its mouth to devour it whole, *Aequorea victoria* has no idea that members of a different species, Homo sapiens, have imitated its hunting technique to obtain not one but two Nobel Prizes and shed a completely new light on their understanding of the world.

Imitating the production of this fluorescent jellyfish protein, known as GFP (green fluorescent protein), by synthesizing it in a laboratory has revolutionized modern biochemistry.

Proteins are the expression of genes. The DNA code is translated into them, every combination of four "letters" of the DNA code translating into an amino acid in the protein chain. They then code all the internal orders and mechanisms of living organisms. Using DNA, we now know how to synthesize proteins in vitro, and even assemblages of various proteins, some grafted onto others. By grafting the jellyfish's GFP proteins onto other proteins, we can endow them with extreme fluorescence, which we can then follow inside cells to visualize the way they function, without interfering with the living mechanisms. Biologists have thus been able to observe neurons as they communicate without disturbing them, genes as they are translated, and many other secret processes of living things, simply by making them glow. Thanks to GFP, we can now visualize and understand mechanisms that are invisible but fundamental to our health. The pioneers of this discovery received the Nobel Prize for chemistry in 2008.

To better observe these fluorescent proteins, a new generation of microscope has been developed, the "super-resolution microscope," which is capable of seeing objects that should be invisible because they're smaller than the wavelength of light itself. This scientific breakthrough earned a 2014 Nobel Prize for its inventors: Osamu Shimomura, Martin Chalfie, Roger Tsien, Eric Betzig, William Moerner, and Stefan W. Hell. These physicists and chemists of various nationalities—American, German, and Japanese—and of equally varied descent— Romanian, Chinese, Japanese, and American—succeeded in working together to change our lives, even though the general public has never heard of them or their discoveries. Like the *Aequorea victoria* jellyfish in the deep, these inventors and their illuminations glow in the darkness.

Just as imitating the techniques of underwater creatures can help advance technology, taking inspiration from their ways of life might help improve our society. The legend of the mirror posited the existence of an underwater civilization, but there are actually plenty of creatures living under the sea whose communities could serve as models for us!

In fact, the sea is host to dynamics from which our world could certainly benefit. There is no waste in the aquatic eco-system. Our cities could learn from the optimization of space in coral reefs. The way marine species live side by side could serve as an example for human society. Decision-making in a school of fish, synchronized without any leaders, could even inspire new political ideas.

But why wait for the inventors or for the urban planners to act on these examples? Even on an individual scale, marine organisms and their astonishing lives have useful ideas for us all.

Like the coral that cultivates within itself a community of algae and bacteria, we could cultivate within ourselves the examples of life that the sea offers us. We could take inspiration from the perseverance of the eel, which never abandons its goal of reaching the ocean, and whose optimistic hope allows it to live forever. Or we might emulate the creativity of the oyster, which, wounded by a grain of sand wedged in its shell, turns this problem on its head and, eventually, into a pearl.

Perhaps the medieval scholars and Latin poets weren't wrong to cling so stubbornly to their legend of the mirror. What if we tried one last time to believe it, as they did back then? What if we gazed behind the mirror and sought to determine who, in our land-based world, would be equivalent to the marine species we already know?

Maybe we would notice that certain people around us are the terrestrial counterparts of the animals we encounter at the bottom of the sea. You'll certainly recognize, among your friends, the "sardines," who prefer the safety of their shoal and shy away when isolated. You probably already rub shoulders with mirror images of the scallop, people who, without wanting to speak directly, still tell us plenty about themselves. And we all know octopus equivalents, capable of adapting and feeling at ease in all situations, talking with their hands and transforming themselves depending on whom they're speaking to. And then there are those whose rhetoric is like the universe of the sole, flat and two-dimensional, while conversely, there are others who venture in three dimensions into the deep, varying their gestures and intonations.

Look closely. You might just find in the crowd a few shellfish, who are discreet but nevertheless have many stories under their shells for anyone willing to listen. Or the giant oarfish from the sea serpent legend, whose reputation is built on rollicking stories, and whose everyday reality might be no less wonderful for being subtle and understated. Or else the prawns with polarized colors, the electric fish, or the ultraviolet-colored creatures . . . those who see the world in a way we don't, or are decked out in colors invisible to our eyes.

With luck you might even cross paths with the lone whale who sings even though no other whale can ever hear him. Who knows, maybe some among us might even finally be able to answer his call?

AQUATIC DIALOGUES

In which we remember the remora.

In which we hark back to the friendship of the killer whales of Eden.

In which dolphins help humans, and make fun of them.

The sea talks to us. Why not answer it?

The comber was the first fish to show me the subtle magic of conversation with undersea creatures. This close relative of the grouper, which largely resembles a perch, lives on the rocky seabeds of the Mediterranean. It is a very curious sort of sentinel. When an outsider arrives, the comber approaches and warns all the other fish. It's often the comber that helps us locate octopuses. Each time I passed a rock where a comber lived, it would emerge from its hole and plant itself opposite me, calmly swimming in place and

looking me right in the eye, intrigued. This was not a mind-
less reflex. It was trying to determine what lay concealed behind this strange new creature in a diving mask and wetsuit.
We shared our mutual curiosity through signs and glances.
It was a very rudimentary exchange, but an exchange nonetheless. Neither of us could understand everything the other
wanted to tell us. Still, it's not necessary in any dialogue to
understand every single thing. It's not even possible.

I've been lucky enough to cross paths and share astonishing encounters with underwater species. I'll never forget the
innocent curiosity of the creatures of the open water. Completely wild, they are unfamiliar with human beings, and
generally not afraid of us. Their first reflex is always to come
and find out all they can about us. Locking eyes with an
ocean sunfish is a strange sensation. Far from our shores,
this fish—roughly two yards in diameter, flat and gray like
a flying saucer—will spontaneously swim toward the boat
and turn on its side in order to observe the occupants. It's
an unknown, solitary creature, so immensely unlike us, who
takes an interest in you, and who clearly wishes to learn
your story. A rare occurrence in an indifferent society such
as ours. Mobula rays, on the other hand, seem frightened at
first by passing boats. But as soon as the engine slows, they
make a large half turn beneath the hull and draw closer, cir-
cling in a strange dance. Their huge triangular wings, white
on one side and black on the other, twirl as they refract the
rays of light in the blue. Their round eyes stare at the ship's

gunwales, where astonishing bipeds blurred by the water's rippling surface briefly upend the routine of day-to-day pelagic life. Families of pilot whales, large black cetaceans that migrate through the French Riviera in summer, may spend hours frolicking around a boat. Often these animals engage in "spy hopping," sticking their heads out of the water to better observe the open-air world and its inhabitants.

Cetaceans love getting a glimpse into our environment and observing us. When a humpback whale in the Pacific raises its eye out of the water to see us better, and tries to make gestures out of the water by waving its long pectoral fins in the air to gauge our reaction, it's clear just how eager these creatures are to interact with us.

This dialogue is a lost art. In all likelihood no one has ever had a real conversation with marine animals the way we human beings do with each other. But no doubt many of our ancestors mastered certain aspects of such a dialogue, back when their existence was indissociable from natural ecosystems. A few snippets of those interactions have survived into the present and serve as evidence of the possibility that we might one day reestablish this kind of contact.

The Australian Indigenous civilization has endured for forty thousand years. This people has had plenty of time to weave a close and thoroughly mysterious relationship with nature. Among the enigmas of their long-forgotten techniques

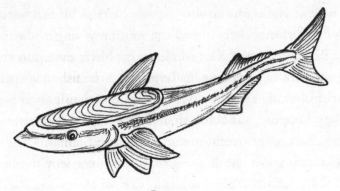

REMORA

is the way they were able to converse with the remora, that aforementioned suckerfish that Pliny suspected could slow down ships.

After Australia was "discovered" by Europeans, several explorers described a novel fishing technique practiced by the Indigenous people of the Torres Strait. In order to capture tortoises, sharks, and large fish, the islanders availed themselves of help from remora attached to the end of a line. The fishermen would slowly approach their intended prey in a canoe filled halfway with water, in which the remoras were kept nice and wet, attached to the bottom of the boat by their dorsal suckers. As soon as they spotted a turtle or shark, the Indigenous people would pull the remora from inside the hull and carefully lower it over the side. The remora would then swim discreetly along, gaining the trust of the shark or turtle, and attaching itself to the larger animal with its suction cup as remoras ordinarily do in the wild. The Indigenous people would then begin pulling the line tighter. The remora would not release its grip. On the contrary, it would back up to make its sucker cling even harder to the prey, which was thus brought to the boat. Some English explorers even reported that the remora tugged on the line to alert the fisherman, as if by telegraph, that he should give the line some slack because the captured fish was about to dive violently into the depths. The complicity between human beings and remoras was so great that even if the line broke, the remora would generally swim back and latch onto the boat. Between outings the remora was kept in a basin full of clean water and fed daily. This is how the fishermen managed to catch turtles, sharks, and a wide assortment of large fish. Their traditional fishing methods never endangered their natural resources. Aboriginal traditions naturally required fishing quotas, setting aside the consumption of each species for a different phase of life. The flesh of large marine creatures was therefore a privilege accorded only to the elderly. The tribes thus avoided overfishing these species, which reproduced slowly, and also prevented mercury poisoning. Mercury, which accumulates in the bodies of large predators, is especially harmful to children and pregnant women.

Explorers' accounts of fishing with remora seemed too fanciful to be true, at least to the intelligentsia of big cities. And yet travelers continued to return with descriptions of essentially the same technique, accompanied by copious illustrations and details. What's more, this technique was also observed outside Australia, in the farthest reaches of the globe. Christopher Columbus was the first to mention it, during his voyage to what he believed to be the East Indies. Similar techniques were reported throughout the Caribbean, from Cuba to Jamaica. French naturalist Philibert Commerson observed the practice in Mozambique in 1770, and the British consul Frederick Holmwood reported it from Zanzibar in 1881. But the populations that possessed this body of knowledge gradually died out; their cultures and traditions vanished upon contact with the West.

In 1905 the American naturalist Charles Frederick Holder decided to try using a remora to capture a turtle or a shark, hoping to settle for mainland skeptics whether it was possible. Basing his effort on various observations and technical descriptions, he tried his luck in the coral reefs of Cuba. With each attempt the remora did exactly as it pleased, and not as Holder desired. It would either refuse to swim toward the intended prey, or latch on with its sucker but release its grip at the slightest tug on the line, or else begin to flee in a way that whetted the shark's appetite, so that it was devoured in a single gulp. The experiment was a fiasco. Holder concluded that the Indigenous people and other civilizations no doubt possessed secrets that allowed them to fish with remora, which secrets surely had to do with convincing the remora to collaborate, and with fastening the line to it without making it feel restricted. He suggested gathering more information about the techniques in question before attempting the experiment a second time.

But no one ever seemed to find the opportunity to do so. The art of fishing with a remora, quite a complicated tradition to implement, was lost with the development of more modern techniques. Ethnologists observed the method in practice among very remote tribes as recently as the 1980s, but none of those observers were able to understand, much less describe the secret of conversing with a remora, how the tribespeople asked for the fish's help, or how they gained its trust. The secret was surely concealed somewhere amid the multiple rites that governed the act of fishing, hidden in a magical chant or traditional dance, and handed down as story, solely through oral tradition. Now there is no one alive who can speak to remoras.

The Indigenous people were not the only Australians to converse with marine animals. For more than a century, the English colony of Eden in New South Wales, located in southeastern Australia, was the setting for an extraordinary friendship between humans and killer whales.

It was no doubt the Aboriginal workers from the Yuin tribe who taught English whalers how to talk to killer whales. In the 1860s Australian whalers hunted the humpback whale from modest rowboats with hand-wielded harpoons. It was a dangerous profession, but necessary for subsistence in these isolated regions. Alexander Davidson and his son John, experts at ship repairing, had decided to take a stab at this adventurous pursuit.

The Davidson family was quite insistent about its Protestant morality and values. Convinced that equal work deserved equal pay, they paid their Aboriginal employees the same salaries as their white employees, an exceptional policy for the time. They thus won the gratitude and respect of the Yuin, who in return taught them how to solicit the help of killer whales in hunting humpback whales. Thus the Davidsons established an alliance with killer whales that made them the whaling experts of the port of Eden.

The killer whales patrolled along the coast, and if they happened to see humpback whales passing by, they would slap the water's surface with their tail flukes to alert the whalers. From shore, the inhabitants of Eden would hear and see giant explosions of spray in the water, and row their boats out to sea at top speed. The pod of killer whales would escort and guide the harpooners while pushing the target whale in their direction. The humans and killer whales had developed signals, in the form of either oars or tails slamming down on the water's surface, and used them to communicate, directing the course and strategy of the hunt. The condition of this alliance was unswerving respect for the "law of the tongue"—the hunters were bound to give the killer whales the humpback whale's tongue, a choice morsel, as a reward. This established a genuine complicity between the men of Eden and the pod of killer whales, one that went well beyond a mere quid pro quo. Each killer whale had a name and a personality. There was an especially strong friendship between Old Tom, a particularly charismatic male killer whale, and George, the Davidsons' youngest son.

Old Tom had been assigned by his fellow orcas to alert

the humans, acting as an intermediary between the two species. The whalers nicknamed him "the humorist," for his numerous pranks. He liked to latch onto the lines of the boats,
clamping down on the ropes with his teeth and letting the
oarsmen drag him through the water. Slowing them down
this way was a great source of amusement for Old Tom, who
never tired of playing tug-of-war. But when the time came
to pursue a humpback, Old Tom was all business. He would
haul the boats out himself, taking the ropes in his jaws and
towing them toward the cetaceans, allowing the oarsmen to
save their strength. As a result, some of his teeth were badly
worn. If a sailor fell into the water, Old Tom would swim
to his rescue, keeping him afloat and protecting him from
sharks. George Davidson swam regularly with Old Tom for
the simple pleasure of it; to him, the killer whale was a mem-
ber of the family. The killer whales guarded his crew, and in
return George guarded the killer whales. He advocated for
laws sanctioning their protection, reported the Norwegian
whaling crews who hunted them to the police, and came to
their aid to free them when they got entangled in netting.
This friendship between humans and killer whales lasted for
three generations, from 1840 to 1930, and it was documented
by invaluable eyewitness accounts, films, and photographs.
While the rest of the world was busy developing motorboats
and explosive harpoons to decimate whale populations on
an industrial scale, back at the port of Eden, humans continued cultivating their longstanding friendship with the killer
whales. They hunted whales only in rowboats and took only
as much as was needed to ensure the colony's survival.

OLD TOM

Woe betide he who betrays the sea. In 1930 the hunting season was very poor, the Norwegian industrial whalers having
let only a few whales escape to Eden. A farmer named John
Logan was working as a harpooner aboard George Davidson's
boat the day Old Tom managed to force a small whale to
the surface to be captured. The whale was relatively scrawny,
probably the last of the season. When the time came for the
killer whales to claim their reward, Davidson and Logan had
a disagreement. Logan thought the whale was too small to
share; it wouldn't even provide enough oil for the lanterns to
last the winter. George, on the other hand, cared deeply about
respecting the immutable law of the tongue, as his parents
and grandparents had done before him, and the Indigenous
people long before any of them. There was a storm on the
horizon, and it was time to get back to port. Losing patience
with the discussion, Logan ordered the men to haul the whale
to shore. George couldn't insist past a certain point, given the
crew's opposition. Old Tom followed the boat in disbelief, at
first taking its departure as some sort of jest. He tried to tug at

the whale, then to slow the boat by dragging on its lines. But the crew only pulled harder toward harbor. This was their last game of tug-of-war, with a sad finale. Old Tom lost several teeth, and watched as his share of the whale was torn violently away. Logan's daughter, who was on board that day, reports that when the injured killer whale swam back to the depths in disappointment, her father murmured, "My God, what have I done?"

The killer whales never came back to help the whalers of Eden. Without their assistance, the inhabitants of the port never caught a single whale again.

A few months after this betrayal, sailors found Old Tom's body washed ashore in a nearby bay. The loss of teeth had no doubt condemned this already extremely aged creature to death by starvation. A guilt-ridden Logan paid for the construction of the chapel where you can still go today to see Old Tom's skeleton, among other souvenirs of this former alliance with the ocean. The port still exists. It lives up to its name—Eden, lost paradise of a friendship betrayed.

Some of these traditional partnerships between humans and the sea persist to this day.

Cooperation between humans and dolphins is nothing new. Pliny the Elder wrote about it. In a lagoon connected to the sea that he called Latera, near the present-day town of Palavas-les-Flots on the French Mediterranean coast, the inhabitants developed an astonishing friendship with the dolphins. During the annual migration of the mullets, "the town's entire population" gathered on the beach to call the dolphins, chanting the name of "Simo, Simo," into the north wind. According to Pliny, this name evoked the Latin word *Simius,* meaning "flat nose," and the dolphins, possessed of a self-deprecating sense of humor, recognized themselves in the call and, amused by the affectionate mocking of their noses, swam to shore. So the large bottlenose dolphins would appear along the beach in vast numbers, pushing mullets into the men's nets and taking advantage of the wall formed by the nets to devour a few themselves. This sounds like yet another of Pliny's many fanciful stories, and yet there's reason to believe it. Something similar still takes place today. In Mauritania the same mullet-catching technique is used by the Imraguen tribe. These former Moorish slaves, only liberated some decades ago, were required for centuries to pay their masters a sizable tribute in the form of fish. But even in their distress the Imraguen could always count on their allies the dolphins, faithful as they were unlikely. The Imraguen don't call their dolphins "Simo," as the people of the Camargue did in Pliny's time, but they do slap the water with a very specific rhythm, playing a sort of percussive tune to lure them ashore. Then they work with the dolphins to drive the leaping mullets onto the beach and into a maze of nets. Unfortunately this technique is gradually falling out of use with the rise of more "modern" ways of fishing, and is now practiced only rarely in Mauritania.

In Brazil, however, in the village of Laguna, roughly two

hundred fishermen currently live in symbiosis with a family of dolphins. In their muddy lagoons, they fish for mullets with a casting net, a bell-shaped net that they toss over the fish. The dolphins can see the mullets in the muddy water thanks to their sonar, while the fishermen can't. On the other hand, the fishermen can catch the mullets in their nets, while the dolphins can't. No one knows today when this alliance between humans and dolphins in Laguna first began, but the symbiotic relationship has become a vital source of livelihood for both species.

The dolphins have even invented an entire language to communicate with the fishermen, guiding them to the best spot to cast their nets with a gesture of the head or tail. This language is passed down from one generation to the next. Even more interestingly, this group of dolphins has developed its own cultural traits. They prefer to keep to themselves rather than mix with other dolphins. Acoustic recordings of their sounds reveal that they have their own "accent," their own special whistles that clearly differentiate them from dolphins of the same species who don't "talk" to humans. The fishermen also have their own expressions, a kind of jargon linked to their interactions with the dolphins. They can recognize each dolphin and have given them names. Underwater the dolphins have names for each other as well. These are two distinct cultures joined together on the sandbars of Laguna—two stories, on and under the water, being written together, in the Portuguese of the fishermen and the whistles of the dolphins.

Relations between humans and sea creatures don't necessarily entail a joint quest for food. Often they're simply the product of reciprocal curiosity, without any ulterior motives.

While diving in Polynesia's Tiputa Pass, I had the good fortune of observing bottlenose dolphins underwater, and witnessed an incredible exchange. These dolphins naturally come to interact with divers as a form of play. They're mischievous and affectionate, so eager for physical contact that you sometimes have to feign indifference and resist the urge to pet them so as to ensure they remain wild and free. The appearance of a dolphin underwater is surreal. It's as though you're looking at a toy animal or watching a special effect in a movie, so strange and so perfect are these creatures.

Around the air bubbles that rose to the surface from the divers, two bottlenose dolphins came and went, intrigued. One was watching us out of the corner of its eye. Calmly, with an air of complicity, it circled and darted through the curtain of bubbles and then stopped, letting itself sink back as if weightless. Then it began to make tiny movements with its fins, reclining clumsily on its back and blowing large streams of bubbles through the blowhole on top of its head. As my eyes met its mocking gaze, I experienced the simple joy of someone in on the joke: I suddenly realized the dolphin was imitating us. It had seen us swimming clumsily in midwater and emitting bubbles, and it was trying to do as we did but with a hint of exaggeration. Was this an attempt to learn from us, some innate instinct to imitate, or was it simply a form of amused mockery? The answer was likely in its mysterious whistles, which we were unable to decipher. But even without understanding each other's language, we shared a moment of real complicity through this game of imitation. This time it was the ocean that saw its reflection in us, as if in a mirror.

True dialogues with marine creatures, conversations that are sincere and lasting, never require that we understand their language, nor that we teach them our own. That's not to say this sort of teaching is not, to some extent, possible. We've managed to teach dolphins and seals in captivity to recognize, and thus react to, a considerable number of "words." But in such a case, we human beings are only trying to make ourselves heard, not start a conversation.

The Imraguen, Indigenous people of Australia, and the divers of Tiputa don't try to exchange words, only to share that which transcends words. They don't approach animals to teach them how to speak; instead they try to become part of the animals' world. They interpret the sounds of their counterpart without understanding them completely, but intent overcomes any language barriers. We can hope that one day science will decipher the languages of undersea creatures. Perhaps science can even make our language accessible to them, allowing us to translate our conversations in both directions. But we know such a translation isn't necessary to talk to them. Some humans have known how to do it for thousands of years.

These wordless conversations are inspiring, and might even be taken as an example of how humans can learn to speak to one another. Each person has a language all their own, a bit like the dolphins and the fishermen, and others can never completely decipher it. We try too hard to be understood when we speak. We try to speak the language of the other person, or to force the other person to understand our own language. But what if we all expressed ourselves freely, naturally, in our own way, with our own personal style? What if we simply listened to others with our heart, without trying to translate every little word, and spoke that way too, without fear of being slightly misunderstood? Dolphins speak their dolphin language, humans their human language. All the same, in the currents of Polynesian reefs, they hear and understand each other.

Seabirds apply this principle, at the top of their lungs and over the open sea, as they hunt for anchovy. The tern, for instance, squawks whenever it spots something interesting, thereby alerting all other creatures to it. Be they gulls or shearwaters, boats or whales, they all hear its voice and its signal. Little does it matter if no one understands what they're saying. This small white bird, capable of migrating from one pole to the other, is so charismatic it can convince any animal to follow it with its shrill cries. I discovered this fascinating dialogue with these birds when fate brought me into contact with another species, a marvelous and colorful character—the bluefin tuna.

IN TUNE WITH
THE TUNA

In which we can interpret the flight of birds.

In which canned tuna is transformed into canned music.

In which stories are told in all tunas.

The tuna entered my life in likely the same way it entered yours—between two slices of white bread, or sprung free from its can to join in the school-cafeteria salad. In other words, as "flake" tuna.

Years later I encountered in person the formidable creature that gives us these "flakes."

Halfway between the European mainland and Corsica, you're entirely encircled by the horizon. The endless 360-degree expanse of the open sea is too much for some people. The blue depths below can also strike fear into our hearts. It's a legitimate cause for vertigo. The boat floats suspended above a two-thousand-meter drop, hovering over abrupt mountains and deep canyons—all hidden behind an infinite blue gulf of immensity.

But I've always found the solitude of the open sea rather comforting. Aboard a ship, in the midst of that flat expanse, you can see everything coming from a long way off.

Dawn was spreading across the sky, scattering the morning's orange curtains into a pervasive blue. We scanned the ripples of the sea with squinted eyes. To the east the water scorched our gaze like a mirror, as if the sun had trickled onto its surface in an incandescent puddle. In the other direction was an intense and soothing indigo.

At first there was nothing in sight. Calm expanses, more ripples. From time to time, the occasional tipsy wave. Just water and air.

"Is there something over there?"

A tiny dot had just moved through the cross-eyed view of my binoculars.

"Where?"

"Five o'clock."

"Yes. A bird, I think . . ."

A small white spot had appeared out of the blue. It's astonishing how suddenly birds appear at sea. Perched on the water's surface, in the hollows between waves, they escape even the most finely calibrated binoculars. And then all at once there they are, at the edge of the sky.

The bird was flying straight ahead, looking determined.

"Looks like a tern. Let's follow it."

Soon there were two terns, then ten, all sprung out of nowhere as if by magic, flying on the same course. They called to each other loudly, confidently.

Then shearwaters appeared, tacking upwind along the surface. And countless gulls filled the sky, whirling around in their gangly flight, as if we had just tossed a pinch of salt and pepper into the air.

The terns climbed higher, soaring and diving in a frenzy. Suddenly one turned around, its tail fanned out, and made an abrupt about-face. At the other end of the horizon, a thin white line was starting to take shape. The tern had spotted it. Alongside the flock of birds rushing along amid loud cries, we followed the tern's path. And the sea began to boil.

In the blink of an eye, the sea was no more than an expanse of churning foam for hundreds of yards, an immense lapping, a dazzling chaos. Thousands of sparkling anchovy rose to the surface, panic-stricken, assailed by predators. Northern gannets dive-bombed the churn like falling javelins. Dolphins darted deftly in and out, slicing through the swirling waters. Terns dove in frenetic flocks, shearwaters belly-flopped into the water with childlike giddiness. I had only just glimpsed the gleam of massive, dark, streamlined shapes, hurtling through the air and slapping the waves as they fell in an iridescent spray, when suddenly the air echoed with the terrible, deafening, formidable cry of the tuna.

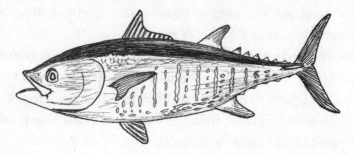

BLUEFIN TUNA

The tuna's cry comes from the mists of time. It has rung in the ears of the people of the Mediterranean for five thousand years, taking on different timbres and tones through the ages.

For our ancestors in the Neolithic era, the cry of the tuna was one long note, blown into a large conch shell by the tribe's lookout, who would wait for weeks at a time, scanning the waves from high atop the cliffs, watching for the annual gathering of migrating tunas to pass between the rocks and the beach. Each night of his relentless dreaming and waiting, the lookout would recall so many legends about these luminous creatures and their titanic strength that he wound up forgetting what they even looked like. To remind himself, he'd go to the caverns where the ancients had drawn them in charcoal. And when he finally saw them appear in full color, in the transparent waters of the narrow, steep-walled inlets, he'd blow with all his might into the immense mother-of-pearl conch shell.

Archaeological digs tell us that Neolithic tribes caught tuna off rocky capes, where schools of the big fish swam along the coast and could be encircled and driven to beach themselves on the shore. In Provence, Sicily, and Crete, cave paintings testify to the animal's spiritual importance, and to the weapon the prehistoric harpooners used to catch them. To succeed in capturing a tuna with the resources of prehistoric times was doubtless an exploit that gave rise to its fair share of legends.

The bluefin tuna possesses extraordinary strength. This animal has adapted to life on the high seas, in the swaying, blue solitude.

A tuna has no shelter or place to rest, so it never stops swimming. It lives like a tireless traveler, always following the currents. Even when it sleeps, it keeps swimming. If a tuna stops swimming, it sinks and drowns, because it can breathe only when it is moving forward. Its gills don't function unless a stream of water constantly surges through them. Its body is essentially one giant muscle, fueled by an enormous heart. All the other organs are reduced to the bare minimum. To power that muscular mass, its respiratory and vascular systems are the most efficient in the entire animal kingdom.

But in order to maintain the energy indispensable to its nomadic way of life, the tuna must also feed incessantly. Shoals of anchovy, krill, sardines, and mackerel—all grist for the mill. If necessary, a tuna will even eat jellyfish, in which case it must consume the equivalent of its own weight daily to power its permanent state of swimming. It eats such vast quantities of jellyfish that the tuna population is an important factor in whether a given year is a "jellyfish year" along our Mediterranean coastlines. With such an appetite, the young tuna grows quickly. During its youth a tuna doubles in weight every year.

In its first year of life, a tuna can cross the Atlantic in sixty days. And in order to travel from the warm waters of the Bahamas to the icy seas of Iceland, it is capable of raising its body temperature above the surrounding water temperature. It's one of the only warm-blooded fish on earth.

"A wonder of nature," according to a certain Aristotle, who didn't even know the full extent of it.

The tuna's cry also resounded for the Greeks and the Romans, refined connoisseurs of the animal, which they transported in amphoras to all the ports of the ancient world. They savored the tuna's flesh, aged for years, in olive oil. Back then, tuna was so abundant that it was said that Alexander the Great's fleet had been forced to deploy in battle formation to take on an immense school that was keeping it from following its course.

The greatest minds of the time attempted to decipher the mysteries of the tuna's migrations. They tried to understand why this immense traveler, which vanished beyond the borders of the known world, invariably came back, faithful to the same migratory routes. Aristotle was convinced that it was blind in its left eye and always watched the coast with its right, so as to follow the contour of the Mediterranean coast. He also believed the tuna was frightened by the gleam of certain white cliffs at the entrance to the Black Sea and shifted the course of its migration accordingly.

Our knowledge of tuna has certainly evolved since Aristotle, but the itinerary of its travels remains a vast mystery.

In the Middle Ages the tuna's cry became a song. The song written for the men who worked in the almadraba fisheries. These large labyrinths of nets stretched from shore out into the waters, trapping groups of tuna that had lost their way and delivering them to the mercy of the harpooners. It was dangerous work, wading, armed with a hook, into the middle of a school of frenzied tuna, amid foam and blood, and trying to single out and drive aground an immense animal big enough to feed several families. In order to steel themselves to enter the nets, the men sang together, in chorus and in canon. The almadraba technique, also known as madrague or mattanza, was developed and fine-tuned by all the peoples of the Mediterranean. Each civilization contributed a detail to the technique, adding a refrain of its own to the song. To this day the songs of the almadrabas blend invocations of the Bible and the Quran, Latin superstitions and Iberian legends, in a melting pot of languages spanning the entire Mediterranean.

The tuna has sounded its cry and its songs across the Mediterranean basin for thousands of years. But one day it almost fell silent.

Prior to the 1980s the Japanese didn't have much of a taste for tuna. All the tuna accidentally caught in Japan were being ground into cat food. You can still find old-school sushi connoisseurs in the Land of the Rising Sun who shun fatty fish. To their palates, real sushi is made from sole or scallops.

Unfortunately the shipping companies that exported Japanese technology to Europe and America wanted something to import on their return trip.

It wasn't hard to start a trend, in a country in the throes of an economic boom. All it took was to soak the tuna in water to rid it of that iron taste the Japanese so detested. Thanks to a lot of advertising, a fish that would have been snubbed by Japanese cats thirty years earlier was soon sold for the price of a sports car each to self-proclaimed connoisseurs.

To fuel this lucrative trade, the tuna purse-seiners, huge factory ships loaded with electronics and EU subsidies, were chartered from Europe to attack the spawning grounds of the bluefin tuna. No more almadrabas, coastal seine fishing, line fishing, harpoons. The countless specific techniques of tuna fishing and their age-old traditions were progressively banished from the map, and, in some cases, even outlawed. Tuna had become a private resource, controlled by a few industrial ship-owners. These animals that once fascinated the peoples of the world were now listed on the stock market—all this before they were even rounded up in entire spawning shoals and transferred to gigantic fattening cages far from prying eyes. Then the tuna were shipped in refrigerated aircraft to their final home between sticky rice and soy sauce. The fish's population rapidly diminished, and the rarer the tuna became, the higher its price sky-rocketed. This only incited the seiners to fish more and more, spurring the proliferation of a vast illegal fishing network.

At the turn of the twenty-first century, after ten years of intensive fishing, less than 15 percent of the bluefin tuna population remained.

Who knows by what miracle I was able to hear the tuna's cry once again, that day in the high seas of the Mediterranean. The tuna's return was one of those marvels through which the sea proves to us, with a certain panache, its massive and practically mocking immensity.

The emergency regulations and controls imposed on the seiners at the end of the 2010s certainly helped, but they weren't enough. The Libyan revolution, by sidelining the chief bluefin-tuna-laundering ally of the French seiners, has also played an inadvertent role saving the species. But above all, there are natural twenty-year cycles in the abundance of bluefin tuna that depend on the activity of the sun, marine currents, and other parameters that remain largely unexplained. Thanks to this confluence of factors, perhaps with the assistance of one or two of those old spirits the ancients used to invoke, shoals of bluefin tuna have miraculously reappeared on our coasts, once again abundant, if only temporarily reprieved.

When I went out to meet them that day, it was in the hopes of understanding their mystery, and of helping to protect them. I was taking part in a campaign to tag bluefin tuna aboard a boat belonging to the sea fishing federation of Monaco, a founder and funder of the program. Our objective was to get our hands on one of those uncatchable tuna, in order to tag it and delve into its secrets.

The terrible cry of the tuna had sounded; there was panic aboard the ship.

The golden reel screeched as it unspooled its nylon line sixty feet at a time, groaning with a terrible cry under the weight of the animal's violent flight. Orders were shouted at the stern of the boat—reel the other lines in, set up a harness, everyone to your stations! The tuna, about a hundred yards away now, continued its course, with no intention of stopping.

It was now urgent to follow it, to try to recover some of the line that was inexorably unwinding from the reel. It was a game of strength and cunning to convince the beast to turn around and head back to the boat.

At the end of the line, the tuna no doubt barely felt the tiny curved hook jammed into the corner of its mouth, like a bone from one of its usual prey. And the traction I was trying to exert in the opposite direction, pumping as best I could with the fishing rod, hardly seemed to have any effect on its trajectory.

But finally the tuna seemed to grow tired. Turning in wide circles directly beneath the boat, it rose gradually toward the surface. It wasn't ready to surrender, just marginally weakened. It was almost as if it were simply coming to get a better look at the boat, an unbowed pride in its gaze. It wasn't defeated, it was offering us the gift of its capture. The gaze of a tuna is not something you soon forget.

The animal swam there, tied to the side of the boat, improbable in its perfection, gleaming like a new toy fresh out of the box. Electric blue stripes, harmonious striations of color, coppery patches that might belong on a modern painting, and a perfect balance of hydrodynamics. Behind it a dozen other tunas from the same shoal followed in the boat's wake, darting shadows. When a tuna advances boldly in a given direction, the rest of the shoal will follow it confidently, believing it has something in mind. Even when it leads them straight to a fishing boat.

The animal, weighing in at sixty-five pounds, hovered near the boat, regaining its vigor and its hues. This was the moment. Crouching down beside it, I pulled out of the boat hull a pole equipped with a small red plastic arrow bearing a code written in black numbers. A rapid stab of the arrow into its back, a confident snip of the pliers to extract the hook from the corner of its mouth, and the tuna surged away into the blue sea, swimming calmly, its dorsal fin now adorned with a small scarlet strand of spaghetti.

It was like sending a message in a bottle out to sea. The tagged tuna would cover hundreds of miles and perhaps, one day, cross paths with someone else, who would see the little red plastic spaghetti on its back and notice the phone number on it.

Since I began working with the tuna-tagging program and developing it in France, dozens of sport fishermen have launched their own bottles into the sea on the backs of tunas, fascinated by the sheer beauty of the beast. Many hundreds of tunas have swum back out with red spaghetti strands trailing down their backs. Some have already told the story of their journeys. Fish that were tagged in France have been found again all over the planet—in America, in the Adriatic, off the Balearic Islands. Many are still out there swimming right now, waiting to be found, perhaps ten years from now, who knows where. By then they'll have gained another six hundred pounds.

The tuna's migration remains a mystery, as fascinating now as it was in Aristotle's time, but little by little we're beginning to unveil some aspects of the enigma. There are relatively sedentary individuals that come and go between France and neighboring Corsica. Others set off on immense loops, beyond the Strait of Gibraltar and all the way into Canadian waters.

Documenting these long journeys will eventually lead to the international management of tuna populations, allowing us to better protect them. For the tuna that we might think of as French is also Canadian tuna, Spanish, or Moroccan, and only international regulation can protect these long-distance voyagers. Through such international management, countries that have long been calling for a sustainable tuna fishing industry—such as the United States, Canada, Monaco, and Norway—will have more leverage to protect them.

Another positive development is that, through tagging tunas, a few scattered sport fishermen, the last of the Mohicans to hunt these animals with rudimentary methods, helped to revive a dormant passion for tuna. The same passion that inspired our ancestors all the way back to prehistoric times, the passion of millennia-old traditions that birthed legends and festivals in all the ports of the Mediterranean, and that has reestablished a bond, an equilibrium, a dialogue between humans and the forces of nature. Consequently they've also revived the art of interpreting the flight of birds, of optimistically scrutinizing the horizon, of feeling the shiver down your spine when you hear the tuna's cry. Of admiration and reverie for the tuna's inspiring life, spent swimming ever forward.

We have given the tuna back its former voice.

THE TAIL END

To the east the sun was blinding. In the blue panorama of open waters, broad shafts of light tilted from east to west, dancing to the rhythm of the sea's gentle chop. The sardines swam along, scattered in the morning calm, gobbling plankton on all sides. Above them, puddles of sky undulated on the surface. Pastel pink dissolved into the blue of the day.

Below, their shadows, cast to the west, tumbled toward the depths still full of night.

Farther down, the tuna glimpsed these dancing shadows.

Their approach caused a ripple. All at once the shoal of sardines regrouped, organizing itself into a compact and terrified mass.

To become the mirror of the sea—the sardine knew this was the only way to escape the tuna's gaze. To blend into the seascape, to be no more than a reflection. The whole shoal needed to assume the same angle simultaneously, so that the blue of the water would be reflected on all sides by their silvery scales, so that they could not be distinguished from the void of the sea. To not tremble. Above all, to not allow a treacherous gleam of sky to accidentally glint off the edge of a scale, a snatch of light betraying their presence. The sardine floated perfectly straight, disappearing with its whole shoal in an invisible twitch.

Already, however, amid that reflection of sea on skin was the outline of the phalanx of tunas, arranged in implacable ranks. It was too late to rely on illusions. The tuna's triangular eye had spotted the shoal. The sardine saw those long-finned, black streamlined silhouettes materializing, suddenly taking on color, in unison. The tunas had illuminated their electric blue stripes on an ultraviolet wavelength precisely calibrated to foil the sardines' vision. There was a dazzling flash.

The tunas' attack was sudden and brutal. The first one charged like a rocket into the shoal, which parted to dodge it and had no time to close ranks again. The other tunas were already arriving. They rose up on all sides, breaching the surface to build momentum and then slamming down with a deafening smash in the middle of the disoriented sardines. There were more and more of them at every instant—hundreds of ravenous bombshells slamming into the group of sardines.

The shoal didn't succumb to panic. Dazed, deafened, the sardines knew their one hope for survival lay in staying together, within earshot of each other, organized, acting as one fish. Their mass assumed the shapes of strange swirls and arabesques to baffle their attackers, dodging the tunas' assaults by breaking apart and hurtling back together almost instantly, trying to reflect the rays of sunlight in all directions to scramble the tunas' sight.

But the tunas had come a long way, traveling day and night without stopping, tortured by hunger. They changed strategy, pushing the ball of sardines toward the surface, toward that impassable wall of sky.

The sardine flew into the air, flung by the surge of its neighbors leaping frantically to escape a powerful tuna's charge. For a few seconds suspended in that light, dry atmosphere, before falling back down, it saw the immense pandemonium beneath, the sea bubbling as far as the eye could see under the assault of the tunas, the crackling spray of sardines through the air, and the sky, the empty sky suddenly invaded by frenzied flocks of birds. The water had become a raucous vortex of incomprehensible currents. It was impossible for the sardines to hear each other, to organize, in this tumult of blows.

A tern dove on the sardine's left, leaving a trail of hissing bubbles, then swam back to the surface with a full beak. The sardine felt the shock of the water pressure as it dodged the impact. It no longer knew which way to go. The attacks were now coming from the sea and the air at once. A tuna leapt back and landed in a spray of foam. What remained of the shoal of sardines shifted brusquely. The sardine, isolated off to one side, couldn't rejoin the group. A tuna spotted it and rushed after it briefly before wheeling around to dive, mouth open wide, into the shoal fading into the distance. Now the sardine was alone, cut off from the pack. Visible and vulnerable, far from the protective mass of its fellow fish. Its one chance at survival was to swim straight ahead.

Scales, torn off by the millions, glittered in the blue like snowflakes. The sardine raced forward, terrified of being caught. On the silvery mirror of its skin, the tunas' bloody banquet appeared as a reflection that grew ever smaller. What images that mirror had reflected! So many scenes in which the sardine had made itself invisible, scenes whose colors it imprinted on its skin. Swimming with all its might, the sardine recalled the tableaux it had copied on its scales: the games of dolphins, the hulls of large ships, the rocky shores of distant islands, strange sea turtles. So many secrets carried inside. What would become of these stories? It was just a solitary sardine, so vulnerable, already practically condemned to the tuna's gastric juices. A tiny morsel in a vast food chain, scarcely a mouthful for all the predators on its trail. What to do so that all these stories in which it had been steeped don't dissolve into the maelstrom of the ocean's cycles?

In the summer of AD 79, Mt. Vesuvius erupted, burying the Roman towns of Pompeii and Herculaneum in volcanic ash. Pliny the Elder lived nearby, in retirement, far from his Transalpine Gaul. Fascinated by this extraordinary natural event, he was determined to observe it from a closer vantage to learn more. He rushed to where everyone else was fleeing, sailing across the Bay of Naples in a ship filled with wax tablets to describe the volcanic eruption. The ash turned day to night, pumice fell like hail. Undaunted, Pliny noted down every detail. But once extremely close to the danger, and just as he was about to turn back, he remembered that a friend lived in a villa on the volcano's slope, from which there was no escape but by sea. His scientific quest thus became a rescue mission, which took a turn for the worse. Pliny managed to save his friends from the volcano at the last moment, but was unaware of the danger of the gases emitted by the eruption. Alas, he did not survive the toxic fumes. But his stories did, including his final description of the volcano. The plume of smoke rose, he said, like a stone pine tree. In the thirty-seven books of his *Natural History,* all of the stories he had carried with him endured to be written and shared. Even now, two thousand years later, we can still read them. Pliny was a human. His writings have out-lasted volcanoes and time alike. But what about the sardine? What chance of survival do a sardine's stories have?

The sardine swam and swam until it lost track of time. It failed to notice that the water around it was changing color, that its scales no longer reflected the light blue of the open sea but rather the green of seagrass and the ocher of rocks. Exhausted, it wavered. Disorienting waves pushed it toward an unfamiliar element, land. It barely noticed that a green net was lifting it out of the water, that now it was swimming in a plastic pail decorated with starfish. That was when its gaze met the eyes of a stranger, the eyes of a child. And as it set off again, miraculously rescued, toward the freedom and dangers of the open sea, it decided to share a few of its stories with this child and encourage him to come along for the ride.

Like the sardine, it's time for me to set off again. I have so many horizons left to explore, new fish to meet, mysteries to contemplate or understand. So many species to try to protect, so many challenges to face in order to rediscover my place within the equilibrium of ocean, and of life. Most of all, so many things to learn and discover in the stories the sea creatures murmur to me. Perhaps one day our paths will cross again. Perhaps I'll tell those stories to you.

Perhaps you, too, will cross paths with a sardine one day, or a whale, a planktonic copepod or a seagull, that will take you in turn on a journey full of discoveries. Perhaps you'll tell those stories to me.

In the meantime let's allow these stories to lull us and inspire us to invent more, and share others. The world of the seas is like the world of words, a space of freedom that must remain so. Those who would muzzle words, impose rules on expression and speech, are like those who would place barriers on the seas. The ocean belongs to everyone and to no one. The imagination, too. So whether we are a solitary whale who speaks its own language, or one of the anchovy swimming in the formation of a vast shoal, or an inventive octopus or a tenacious remora or a discreet lobster, let's sing our stories, each in our own way.

I hope these aquatic reveries leave you with a few dreams, a few ideas, and the desire to share them with friends. I hope they even give you a new perspective on creatures you never really considered before, and a desire to listen to them, know them, and protect them.

I hope this book has brought you to horizons that are unknown and yet within reach. I hope you will take them with you, the way you might a seashell from the beach. And I hope that now and then you lift that seashell to your ear. They say if you listen closely, you can hear the sea.

EPILOGUE

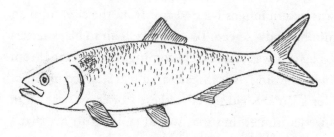

ALLIS SHAD

How hard it is to write a book, especially during the dog days of summer! Your fingers tap the keyboard as if it were a muffled piano. You correct, you erase, and then the computer freezes just when you have a flash of inspiration.

Outside, musicians are playing the trumpet. The same tune, over and over again, every night for the past three weeks. And there's no way to type to their rhythm. Writing on paper is more agreeable, but the scribbles and erasures remain. The page fills up as the aspirin bottle empties out. A heavily marked page is less intimidating than a blank page, but only by a small margin.

Stories are meant to come to life, to be recounted out loud, with sweeping gestures, with friends present, with questions and stunned expressions. It's hard to write them, because you have to pin them down, turn them into portraits, choose a single angle, and reduce them to a single dimension. Stories of the sea are even wilder, even more indomitable. That's probably why no sardine has ever written a book.

I've often wondered what the sardine would think of what I was writing. I was afraid of growing distant myself from the world I was describing, as I stared at it from the city, from an office, from behind a screen. By reducing it all to black letters on white pages, I was afraid of becoming disconnected from these stories, of losing their thread. That fear stopped me from writing, for a time. I had to relive these stories, to hear them again. I needed something that would confirm to me that I hadn't strayed from that world.

A humming sound on the table. A notification made my cell phone buzz. The best way to lose your focus and your inspiration. Distracted, I opened the message. Instagram. A friend was alerting me to an extraordinary event: several allis shad had been spotted in Paris.

To me the allis shad was little more than a legend. Someone

told me when I was young that this giant sardine, much like salmon, once swam from the sea back up all the rivers of France to spawn in headwater streams. The last time allis shad had swum up the Seine was in 1920. After that, with dams and pollution blocking their migration, the fish disappeared and took a thousand gastronomical and folk traditions with them. But as the quality of the river's water gradually improved, the shad had discreetly returned—or such was the rumor circulating online. I couldn't miss their return. I replied to the message: *Tomorrow night, meet by the water.*

On that early summer evening the Seine reflected the sky through a fine veil of insects. The river currents swirled on the surface in decorative arabesques. All at once, they appeared. Silvery gleams shot through the eddies, large half-moon fishtails slapped the surface, and we saw the long backs of bluish, leaping fish. Dozens, maybe even hundreds of shad. They were swimming against the current, driven by the instinct to reproduce upstream. These immense sardines had traveled all the way from the distant Atlantic Ocean and were now streaming through Paris. They were coming back from the depths of an era even our grandparents had never known. They had vanished into the shadows of the abyss for roughly a century, and that night they were back for the first time, as if it were the most natural thing in the world. With all the fresh exuberance and naïve grandiloquence of nature, they were leaping boisterously in the middle of the Seine.

After receiving intelligence from various associations, I had my mission. I was to catch a shad and peel off one of its scales so scientists could retrace its path. And that's what I did.

It was thrilling to hold such a fish in my hands. I gazed for a long time at the shad's golden mask and indigo reflections, then watched it swim away. Its brilliant colors had reflected so many distant landscapes, its gaze held so many memories of the ocean . . . The giant sardine swam away with a swift swish of its tail, continuing upstream toward the river source.

The next day, I returned to my manuscript, reassured and calm. I'd never have believed a sardine would come pay me a visit, that it would even venture from so far away into the heart of the city, right up to my home, to whisper its stories in my ear.